COSMOCHEMISTRY

ASTROPHYSICS AND SPACE SCIENCE LIBRARY

A SERIES OF BOOKS ON THE RECENT DEVELOPMENTS
OF SPACE SCIENCE AND OF GENERAL GEOPHYSICS AND ASTROPHYSICS
PUBLISHED IN CONNECTION WITH THE JOURNAL
SPACE SCIENCE REVIEWS

VOLUME 40

COSMOCHEMISTRY

PROCEEDINGS OF THE SYMPOSIUM ON COSMOCHEMISTRY,
HELD AT THE SMITHSONIAN ASTROPHYSICAL OBSERVATORY,
CAMBRIDGE, MASS., AUGUST 14–16, 1972

Edited by

A. G. W. CAMERON

Center for Astrophysics, Harvard College Observatory, and
Smithsonian Astrophisical Observatory, Cambridge, Mass., U.S.A.

D. REIDEL PUBLISHING COMPANY
DORDRECHT-HOLLAND / BOSTON-U.S.A.

First printing: December 1973

Library of Congress Catalog Card Number 73–88588

ISBN-13:978-94-010-2660-4   e-ISBN-13:978-94-010-2658-1
DOI: 10.1007/978-94-010-2658-1

Published by D. Reidel Publishing Company,
P.O. Box 17, Dordrecht, Holland

Sold and distributed in the U.S.A., Canada, and Mexico
by D. Reidel Publishing Company, Inc.
306 Dartmouth Street, Boston,
Mass. 02116, U.S.A.

TABLE OF CONTENTS

INTRODUCTION

The International Association of Geochemistry and Cosmochemistry was organized in 1967, and held its first meeting at UNESCO Headquarters that year in association with its symposium on 'The Origin and Distribution of the Elements'. The Association is a member of the International Union of Geological Sciences, and holds regular meetings at the time of the International Geological Congresses, the last of which was held in Montreal, in August, 1972.

The IAGC was organized to coordinate activities on an international scale in a wide variety of branches of geochemistry. Its activities are carried on through Commissions and Working Groups, and by means of symposia and other international activities. It has national, corporate, and individual members.

One of the first actions taken by the Council of the IAGC when it met in 1967 was to establish an initial set of Working Groups to commence the activity of the organization. Among these Working Groups was one on Extraterrestrial Chemistry, established under the chairmanship of the writer. This Working Group recognized that its basic concern with the chemical composition of cosmic systems was a problem with ramifications in many fields in addition to geochemistry. The other scientific disciplines which are involved include physics, astronomy and astrophysics, and geophysics. The Working Group thus included scientists in these disciplines from the beginning; many of the scientists had already participated in the first symposium of the IAGC. The Working Group has recently been elevated to the status of a Commission.

It has been a basic objective of the Working Group on Extraterrestrial Chemistry to sponsor appropriate international meetings. On behalf of the Working Group, the IAGC has co-sponsored several international meetings, including the meeting on Meteorites, held at the International Atomic Energy Agency in Vienna in 1968. However, the recent International Symposium on Cosmochemistry held at Harvard University on August 14–16, 1972, was the first meeting for which the Working Group on Extraterrestrial Chemistry took prime responsibility for organization. This meeting was a successor to the IAGC-organized symposium on The Origin and Distribution of the Elements held in Paris in 1967.

The 1972 symposium was held the week before the International Geological Congress which met in Montreal. The local hosts for the meeting were the Smithsonian Astrophysical Observatory and the Harvard College Observatory, and local arrangements were ably organized by Dr Henri Mitler.

The International Symposium on Cosmochemistry contained a mixture of invited and contributed papers. A majority of the invited lecturers have prepared review articles based upon their talks, and these are included in the present volume. Also included is a list of invited talks at the symposium, and the contributed papers together with their abstracts.

The papers included in this volume range over most of the major areas of cosmochemistry. The authors include physicists, astronomers, and geochemists. Therefore the present volume should be of interest to people concerned with the interdisciplinary area where these subjects meet in the field of cosmochemistry.

The writer wishes to thank the members of the Executive Committee of the Working Group on Extraterrestrial Chemistry: J. F. Lovering, J. M. Greenberg, P. B. Price, T. C. Owen, B. Mason, D. L. Lambert, and F. L. Whipple, for their help in organizing the papers at the symposium. He is grateful to Prof. Earl Ingerson, then President of the IAGC, for his help in establishing the symposium. Special thanks are due to Dr Henri Mitler, chairman of the local organizing committee, for the excellent way in which he took care of the needs of the visitors, and maintained the smooth operation of the symposium itself.

A. G. W. CAMARON

CONTRIBUTED PAPERS

Metal Abundance in Population I Stars
　　R. C. Henry, J. E. Hesser, and W. McClintock
Nucleochronometers for the p- and s-Processes
　　J. Audouze and D. N. Schramm
Chronology of Galactic Heavy-Element Nucleosynthesis
　　T. P. Kohman and J. M. Huey
^{244}Pu as a Possible Indicator of Interstellar Dust Within the Solar System
　　G. A. Cowan
Nonequilibrium Chemistry and the Composition of Cosmic Clouds
　　B. D. Donn, E. W. Chapelle, W. A. Payne, and L. J. Stief
Amino Acids in Meteorites
　　J. G. Lawless, E. Peterson, and K. A. Kvenvolden
Uranium and Thorium Microdistributions in Meteorites
　　G. Crozaz, D. Burnett, and R. Walker
Some Constraints on Current Outgassing of Water Vapor on Mars
　　S. J. Peale
^{40}Ar/^{39}Ar Ages of Lunar Samples from Hadley Rille and the Apennine Front
　　L. Husain
Lunar Chronology Based on ^{40}Ar/^{39}Ar Ages
　　O. A. Schaeffer
High-Sensitivity and High-Accuracy Analysis of Rare-Earth Elements in Lunar Rocks
　　L.-D. Nguyen, G. Puil, M. de Saint Simon, and Y. Yokoyama
Comparative Lead-Isotope Dating of Different Classes of Meteorites
　　J. M. Huey and T. P. Kohman
The Problem of Composition Studies of Solid Cometary Material
　　P. M. Millman
Interplanetary Dust: A Source of Primitive Matter
　　D. E. Brownlee and P. W. Hodge
Anomalous Geochemical Composition in Spherules as a Criterion of Cosmic Origin
　　T. Grjebine
Magnetic Field in the Primordial Solar Nebula
　　M. W. Rowe, J. M. Herndon, D. E. Watson, and E. E. Larson
The Europium Anomaly and Rare-Earth and Other Abundances in Calcium-Poor Achondrites
　　W. V. Boynton and R. A. Schmitt
A Search for Interstellar Lithium
　　N. Carleton and W. Traub

INVITED PAPERS

Stellar and Solar Abundances
  B. E. J. Pagel, *Royal Greenwich Observatory*
Nucleosynthesis
  J. W. Truran, *Yeshiva University*
Cosmochronology
  D. N. Schramm, *California Institute of Technology*
Interstellar Grains
  J. M. Greenberg, *State University of New York at Albany*
Interstellar Molecules
  W. Klemperer, *Harvard University*
Composition of Cosmic Rays and Solar Particles
  P. B. Price, *University of California at Berkeley*
Composition of the Solar Wind
  A. J. Hundhausen, *High Altitude Observatory*
The Deuterium Problem in the Primitive Solar Nebula
  H. Reeves, *Centre National de la Recherche Scientifique*
Composition of Planetary Atmospheres
  J. S. Lewis, *Massachusetts Institute of Technology*
Structure and Composition of Terrestrial Planets
  D. L. Anderson, *California Institute of Technology*
The Moon as a Planet
  P. W. Gast, *NASA Manned Spacecraft Center*
The Giant Planets
  W. B. Hubbard, *University of Texas at Austin*
Comets and Interplanetary Dust
  A. H. Delsemme, *University of Toledo*
Chemistry of the Solar Nebula
  J. W. Larimer, *Arizona State University*
Evolutionary Processes in Meteorites
  J. A. Wood, *Smithsonian Astrophysical Observatory*

STELLAR AND SOLAR ABUNDANCES

B. E. J. PAGEL

Royal Greenwich Observatory

Abstract. Recent spectroscopic results on stellar and solar abundances are reviewed with special reference to (a) Standard abundance distribution (Sun, hot stars, diffuse nebulae); (b) Abundance peculiarities related to stellar evolution (red giants showing results of H-burning and s-process, peculiar and metallic-lined stars); and (c) Population effects that may be related to the evolution of the Galaxy (correlation between stellar age and metal abundance, differences in details of heavy-element mixture in atmospheric composition of normal stars that have not reached an advanced evolutionary stage).

1. Introduction

At the Paris Symposium in 1967, there were ten papers on solar and stellar abundances. This time there are only two, so that the subject cannot be covered with any attempt at completeness. This article will try to stress aspects of particular interest to the present Symposium in having a bearing on theories of the origin of the elements, with some asides on items of particular astrophysical interest. The subject can be considered under three main headings:

(a) Astrophysical evidence bearing on a standard abundance distribution.

(b) Abundance peculiarities related to stellar evolution.

(c) Population effects in the composition of stars and the interstellar medium bearing on the chemical evolution of our own and other galaxies.

2. Standard Abundance Distribution

This term may be preferable to the traditional 'cosmic abundance distribution' because we do not really know whether the abundance distribution found in our Solar System has a universal character. At the same time, many features of this distribution are reproduced in the majority of the stars we can observe and furthermore it provides the best available standard against which to judge the predictions of nucleosynthesis theory and external samples of cosmic material. Astrophysical studies of the standard abundance distribution have progressed considerably since 1967, but at the same time have not suggested a need for any serious revisions in the excellent table by Cameron (1968), except perhaps to increase his hydrogen abundance by 25% or so to normalise on to recent solar photospheric abundances of Na, Mg, Al, Si and Ca. We discuss first solar abundances of which a selection is given in Table I.

Some comments on the different methods of determining solar abundances may be of interest. Four different techniques are involved:

(i) Photospheric abundances based on Fraunhofer absorption lines of neutral atoms, singly charged ions and diatomic molecules.

(ii) Coronal abundances based on visible forbidden emission lines of very highly ionised species observed from the corona at eclipses.

TABLE 1

Solar and Solar-system abundances normalised to $\log H \equiv 12$ or $\log H \simeq 12$ (a) The lighter elements up to $Z = 30$

	Fraunhofer spectrum					Solar cosmic rays Bertsch et al., (1972) [2]	Corona Visible De Boer et al., (1972) [3]	EUV Dupree (1972) [4]	Solar System Cameron (1968) [5]
	GMA (1960) [1][a]	Lambert	Kiel	Forbidden Lines	Others				
1. H	≡ 12.00	≡ 12.00	≡ 12.00	≡ 12.0	≡ 12.00	10.86	≡ 12.0	≡ 12.0	11.88
2. He									10.78
3. Li	0.96				0.7:[6, 7]				3.11
4. Be	2.36				1.1 [6, 8]				1.30
5. B					2.8 [6]				2.25
6. C	8.72	8.55 [9]		8.6 [10]	8.6 [11]	8.60		8.6	8.59
7. N	7.98	7.93 [9]			7.88 [12]	8.1		8.2	7.85
8. O	8.96	8.77 [9]		8.9 [13]	8.9 [12, 14]	8.85		8.8	8.83
10. Ne						8.05		7.5	7.83
11. Na	6.30	6.18 [15]			6.30 [16]			(6.3)	6.26
12. Mg	7.40	7.48 [17]				7.60		7.5	7.48
13. Al	6.20	6.40 [15]						(6.6)	6.39
14. Si	7.50	7.55 [18]		7.5 [19]		7.3		7.55	7.46
15. P	5.34	5.43 [15]							5.56
16. S	7.30	7.21 [15]		7.2 [20]		6.8:		7.3	7.16
17. Cl		≤ 5.5 [21]	6.36 [22]	6.3: [23]			(5.5)		4.75
18. Ar									6.82
19. K	4.70	5.05 [15]							4.97
20. Ca	6.15	6.33 [17]					6.4		6.33
21. Sc	2.82				3.04 [24]				2.98
22. Ti	4.68								4.82
23. V	3.70					←———			4.41[b]
24. Cr	5.36				5.85 [25]		(6.8)		5.55
25. Mn	4.90				5.42 [26]		(6.9)		5.40
26. Fe	6.57		7.60 [27]	7.5 [28]	7.28 [29]	6.9: ———→	8.0	(7.3)	7.41
27. Co	4.64						(6.0)		4.82
28. Ni	5.91		6.25 [30]	6.3 [31]			6.6		6.12
29. Cu	5.04		4.16 [32]				(4.6)		4.42[b]
30. Zn	4.40	4.42 [33]							4.64

(b) The heavier elements

	Fraunhofer spectrum				Solar System Cameron (1968) [5]	
	GMA (1960) [1]	Lambert	Grevesse	Others		
31. Ga	2.36	2.84 [33]		2.80 [34]	3.12	Ga
32. Ge	3.29	3.3 [33]			3.56	Ge
37. Rb	2.48	2.63 [35]		2.60 [36]	2.23	Rb
38. Sr	2.60	2.82 [17]		2.80 [36]	3.23 b	Sr
39. Y	2.25				2.12	Y
40. Zr	2.23				2.94	Zr
41. Nb	1.95				1.52	Nb
42. Mo	1.90				1.86	Mo
44. Ru	1.43				1.66	Ru
45. Rh	0.78				0.98	Rh
46. Pd	1.21				1.64	Pd
47. Ag	0.14			0.85 [37]	1.16	Ag
48. Cd	1.46	2.1 [33]			1.79	Cd
49. In	1.16	1.71 [33]			0.80	In
50. Sn	1.54	1.7 [33]			2.09	Sn
51. Sb	1.94				1.04	Sb
56. Ba	2.10	1.90 [17]		2.1 [39]	2.13	Ba
57. La			1.8 [38]		1.02	La
58. Ce			1.9 [38]		1.53	Ce
59. Pr			1.6 [38]		0.69	Pr
60. Nd			1.8 [38]		1.35	Nd
62. Sm			1.7 [38]	2.3 [39]	0.82	Sm
63. Eu			0.5 [38]	0.7 [39]	0.42	Eu
64. Gd			1.1 [38]		0.99	Gd
66. Dy			1.1 [38]		1.02	Dy
69. Tm			0.4 [38]		0.00	Tm
70. Yb	1.53		0.8 [38]		0.78	Yb
79. Au				0.7: [40]	0.76	Au
80. Hg		<3.0 [33]	<2.1 [41]		1.34	Hg
81. Tl		≤0.2 [33]			0.72	Tl
82. Pb	1.33	1.90 [33]	1.83 [42]		1.92	Pb
83. Bi			≤0.8 [42]		0.67	Bi
90. Th			0.8 [42]		−0.01 c	Th
92. U			≤0.6 [42]		−0.17 c	U

a Numbers in square brackets are references. For the List of references to Table I see pp. 20-21.

b Figures From Urey (1967) (V=3.91, Cu=4.21, Sr=2.82) are in somewhat better agreement with photospheric values.

c Abundance assumed to have held at birth of Solar System.

(iii) Coronal and upper chromospheric abundances determined from permitted emission lines of numerous different levels of ionisation in the extreme ultra-violet (EUV) and soft X-ray spectrum coming chiefly from the chromosphere-corona transition region where the electron temperature rises steeply with height.

(iv) Charge distribution of energetic nuclei often called 'solar cosmic rays' emitted from intense flares.

The photospheric method is the classic and fundamental one well described by GMA (Goldberg *et al.*, 1960). To determine an element: hydrogen ratio we need (a) observational data, preferably equivalent widths of numerous weak lines, (b) a good model of the temperature structure of the photosphere, (c) a reliable theory of radiative transfer and excitation-ionisation equilibrium, and (d) good oscillator strengths and (in some cases) collision broadening parameters from theoretical or experimental physics. The first requirement is met by those of the more abundant elements that have optical transitions of low excitation energy or form suitable molecules, e.g. C, N, O, Na, Mg, Al, Si, S, Ca, Sr, Ba and the iron group, whereas many other elements are represented by only a few blended features that require detailed spectrum synthesis (and hence knowledge of velocity fields as well as temperature structure) for their interpretation. Much progress has been made recently through the development of high-resolution low-noise solar spectrometry at Kitt Peak, Liège and Oxford. The second requirement of a good model is also adequately met, since there are powerful checks from continuum measurements over a very wide range of wavelengths and different positions on the disk (Gingerich *et al.*, 1971). Any of the numerous models current in the literature will give abundances from weak absorption lines observed at the centre of the disk that agree to better than 0.1 dex; but a few abundances (e.g. of Mg isotopes) depend on measurements in spectra of sunspots for which uncertainties of model are more serious. The requirement of a good theory of line formation is more controversial, since one normally assumes the lines to be formed in local thermodynamic equilibrium (LTE). For most lines this assumption has no firm theoretical basis (Mihalas, 1970) and has been strongly criticised, but within the accuracy of ± 0.1 dex or so it is in fact a very good approximation in typical cases as can be demonstrated by cross-checking with forbidden absorption lines of [C I], [O I], [S I], [Ca II], [Fe II] and [Ni II] for which there are good theoretical grounds for supposing that LTE effectively holds (Pagel, 1971a). Various other consistency checks can also be made and they imply that errors in abundances due to departure from LTE must be quite small relative to the accuracy one can hope to get in the present state of the art. Another controversial point is the Doppler broadening of lines that are strong enough to be saturated so that they occur on the flat part of the curve of growth. Existing work based on laboratory oscillator strengths (Foy, 1972; Yamashita, 1972) implies the existence of a so-called 'microturbulence' giving a Doppler width slightly greater than that predicted from thermal motions alone. This effect is large and undoubtedly real in supergiant stars (e.g. Van Paradijs, 1972), but its existence in the Sun is rather marginal compared to various uncertainties and its hydrodynamical significance is not well understood (cf. Worrall and Wilson,

1972). The difficulty can be circumvented in the case of the Sun by using sufficiently weak lines, but it causes trouble in differential analyses of other stars relative to the Sun.

The skeleton in the cupboard, so far as photospheric abundances are concerned, is undoubtedly the oscillator strengths, especially for elements of the iron group for which no decent theory is available and the majority of experimental data still come from free-burning arcs whose properties are complicated (cf. Takens, 1970), poorly understood and not properly calibrated (Whaling et al., 1969; Garz et al., 1969; Bell and Upson, 1971). The most notorious case, so far at any rate, is that of iron itself, for which until about 1969 there was a glaring discrepancy of about 1 dex between the photospheric abundance deduced from permitted lines of FeI and FeII and a whole lot of other data: coronal abundances both from visible and EUV, meteorites which in the case of type I carbonaceous chondrites otherwise fit the Sun very well (Urey, 1967) and forbidden lines of [FeII] in the photospheric spectrum itself. A similar discrepancy between permitted and forbidden iron lines was found to occur in Arcturus (Gasson and Pagel, 1966). Since then the discrepancy has been reduced to a much more tolerable level by a whole series of better oscillator strength measurements using wall-stabilised arcs (Garz and Kock, 1969; Bridges and Wiese, 1970) and shock tubes (e.g. Wolnick et al., 1970, 1971). The results derived for photospheric iron abundance still range over a factor of about 2 (from 7.6 to 7.3), with some uncertainty about what happens to the stronger lines which are affected by collision damping as well as by oscillator strength (cf. Pagel, 1971b); but however the remaining arguments come out, it seems fair to say two things:

(i) Within the uncertainties that still exist, there are no significant discrepancies between photospheric and other solar abundances, or between solar abundances and those in type I carbonaceous chondrites apart from the obvious case of volatiles and the unstable nuclei of D and Li. This statement applies also to the measurable isotope ratios $^{12}C/^{13}C$ (Hall et al., 1972), $^{24}Mg/^{25}Mg/^{26}Mg$ (Lambert et al., 1971; Boyer et al., 1971) and $^{85}Rb/^{87}Rb$ (Hauge, 1972).

(ii) The assumptions normally made about the formation of solar absorption lines are adequate within the limitations imposed by laboratory data. Further improvement of these data, including collision damping parameters as well as the oscillator strengths themselves, is something greatly to be desired and encouraged.

Coronal abundances deserve a brief comment. Equivalent widths of visible forbidden lines with well-known theoretical oscillator strengths give an accurate estimate of the population of the upper state of the line relative to free electrons (which produce most of the continuum); the difficulties then come in extrapolating to the ground state and to unobserved states of ionisation. Table I shows the results derived by De Boer et al. (1972) from an analysis of Hawaii-Sacramento Peak observations of the eclipse of 1965, May 30, near sunspot minimum; numbers based on uncertain extrapolations are given in brackets. Even when one can sum over several observed states of ionisation, there are some grey areas due to subtle excitation mechanisms like proton impact and EUV resonance radiation. The EUV abundances are taken from a recent analysis by Mrs Dupree (1972) of OSO IV data on a small region of the quiet centre of the solar disk, using a modification of the method invented by Pottasch (1968). The method depends on collisional excitation rate coefficients and ionisation equilibria, but uncertainties tend to be ironed out when many states of ionisation are observed. Abundances based on only one state of ionisation are shown in brackets.

The solar cosmic ray data (cf. Fichtel, 1971) have been collected for long enough to give reasonable statistical accuracy and the question is: how uniformly are the different nuclei accelerated? The excellent agreement with photospheric relative abundances of C:N:O: Mg encourages one to extrapolate to He and Ne which have the same charge-to-mass ratio; but the iron group does not fit, and recent observations suggesting selective acceleration effects at low energies even among nuclei with $e/m = \frac{1}{2}$ (Mogro-Campero and Simpson, 1972; Crawford et al., 1972) imply that we should look upon these abundances with a certain amount of caution. The helium abundance seems, however, to be confirmed by chromospheric and prominence measurements (Hirayama, 1971).

To conclude the discussion on solar abundances, several of the results have changed only a little since GMA: some have been vastly improved, but there is still a great deal to be learned. Progress is now being made with the iron group, but for the heavier elements above the iron group, most results will be poor for a considerable time owing to shortage of both solar and laboratory data.

How universal are the Solar-System abundances? Later on we discuss the variation in the ratio of hydrogen to heavy elements among the cooler stars, but in the present context it is of interest to study hot stars of spectral types B and A, which are much younger than the Sun, and the interstellar medium from which they are being formed, e.g. the Orion nebula. There are various difficulties associated with the analysis of B and A type spectra. Most of the stars are spinning rapidly, making it very difficult to study the weak lines due to most elements other than hydrogen and helium because they are fuzzed out by Doppler broadening. Those that are not spinning rapidly often turn out to be peculiar, with high silicon, low helium and sometimes other spectral features of the most bizarre kind. Consequently only a handful of apparently 'normal' A and B type stars with reasonably sharp lines have been comprehensively analysed. Oscillator strengths generally seem to be reasonably good because mostly the lighter elements (Wiese et al., 1966) are involved; and the data on line broadening, especially for hydrogen and helium, have been greatly improved through the work of H.R. Griem and others. However, certain lines are rather sensitive to the temperature structure of the atmospheres, for which only theoretical models are available, and for the hotter stars (especially those of class O) departures from LTE become very significant, though somewhat amenable to theoretical treatment (Auer and Mihalas, 1972; Mihalas, 1972).

Table II shows results of some different analyses of two 'normal' B stars τ Scorpii (Unsöld, 1942; Hardorp and Scholz, 1970) and ι Herculis (Peters and Aller, 1970; Kodaira and Scholz, 1970). For τ Sco the differences between the pioneering coarse

TABLE II

Abundances in representative hot normal stars compared to
H II regions and the Sun

	τ Scorpii B0V		Herculis B3V		Local H II regions	Sun
	Unsöld (1942)	Hardorp-Scholz (1970)	Peters-Aller (1970)	Kodaira-Scholz (1970)	(Orion Neb., M8, M17) Peimbert-Costero (1969)	
T'_{eff}	28 150	32 000	20 000	20 200		5800
log g	4.9	4.1	4.0	3.75		4.44
H			12.00			
He	11.2	11.0		10.8	11.0	10.9 [c]
C	8.2	8.1	8.8	8.1	8.7	8.6
N	8.6	8.3	8.3	7.7	7.6	7.9
O	9.0	8.7	9.0	8.4	8.8	8.8
Ne	9.0	8.6	8.6	8.6	7.9	7.9 [b]
Mg	7.8	7.5	7.3	7.3		7.5
Al	6.6	6.2	6.5	6.2		6.4
Si	7.8	7.6	7.4	7.2		7.6
S		7.2	7.1	7.2	7.6	7.2
Fe		7.3 [a]	7.6 [a]	7.6 [a]		7.3 to 7.6

[a] from Fe III.
[b] from solar cosmic rays normalised to C, N, O.

analysis by Unsöld and the recent careful fine analysis are small, never exceeding 0.4 dex; for ι Her it is discouraging to see rather greater differences between the two fine analyses. The other two columns show abundances derived mainly from forbidden lines in three diffuse nebulae (Peimbert and Costero, 1969) and abundances in the Sun; further data for the Orion nebula have recently been given by Aller (1972), who has also found similar results in prominent H II regions of two galaxies in the local Group (cf. Osterbrock, 1970).

The basic impression is one of uniformity; note particularly the agreement between the Sun, formed 5×10^9 yr ago, and the sample of present-day interstellar gas. Relative to these two, τ Sco shows signs of low carbon and high nitrogen, which may be significant, but it is clear that these results are not unanimously found in all investigations of B stars. What is found by everybody, ever since Unsöld's pioneering work, is an apparently high abundance of neon, based on Ne II in τ Sco and Ne I in ι Her, and therefore perhaps to be taken seriously. Furthermore, the N and Ne abundances seem to vary in otherwise normal B stars, since Dufton (1972) has found lower abundances of these elements in the B1 Ib supergiant HD 96248 previously noted as having weak nitrogen lines. Dufton actually finds solar abundances of N and Ne in this star and suggests that there must be inhomogeneities in the present-day interstellar medium from which these stars are being formed. This interesting question may be clarified in the near future by satellite studies of interstellar EUV absorption lines arising in H I regions (Herbig, 1970); EUV data having sufficient resolution to study interstellar lines other than those of atomic and molecular hydrogen are now just beginning to become available.

3. Abundance Peculiarities Related to Stellar Evolution

Abundance peculiarities in evolved red giants have played an important part historically in the development of the theory of stellar nucleosynthesis (Burbidge *et al.*, 1957). Prominent examples are the classical carbon stars of spectral types R and N which seem to be rich in nitrogen and poor in oxygen (Querci and Querci, 1970; Thompson *et al.* 1971; Greene *et al.* 1973), resulting in strong CN and C_2 bands, and often show features of ^{13}C-containing molecules as might be expected from hydrogen-burning followed by convective rising of products to the surface although the-true $^{12}C/^{13}C$ ratios are not always well established (Fujita, 1970); S stars, rich in carbon and heavy elements like Zr, Sr and Ba, synthesised by slow neutron capture (e.g. Boesgaard, 1970), and occasionally Tc (Merrill, 1952; Peery, 1971); Ba II stars, somewhat resembling S stars but hotter, which historically played an important role in the elucidation of details of the s-process (Burbidge and Burbidge, 1957; Warner, 1965); CH stars (Wallerstein and Greenstein, 1964; Wallerstein, 1969) which have similar overabundances again, but superposed this time on a low basic heavy-element abundance typical of extreme Population II; and hydrogen-deficient carbon stars like R Cr B and the so-called helium stars which probably differ from one another only in effective temperature and may show the effects of hydrogen and helium burning in the core after loss of the surrounding envelope (Warner, 1967).

During the last few years there has been more work on superficially normal-looking red giants which suggests that some of the effects seen in carbon and Ba II stars also occur in quite a number of other G and K giants. Iben (1965) predicted that a star of 3 solar masses that has passed through the helium-core ignition stage should have the atmospheric N/C ratio increased by a factor of 3 owing to mixing by deep penetration of the surface covenction zone, and some enhancement of nitrogen has indeed been found in the giant α Ser and the supergiant ε Peg by Greene (1969) and in the giant α Boo by Fawell (1970). These effects are somewhat marginal compared to possible uncertainties in spectral analysis, but a very large enhancement of N has been found in the F-type supergiant α Per by Parsons (1967). Another effect that seems to be more convincingly real is an enhancement of ^{13}C in Arcturus, where the $^{12}C/^{13}C$ ratio deduced from the 2μ CO bands (Greene 1970; corrected for line saturation effects by Fawell, 1970) is somewhere about 10 compared to about 100 in the Solar System. However, the $^{12}C/^{13}C$ ratio of 4-5 found by Johnson and Méndez (1970) in several K and M stars (and a similar ratio reported for α Ori by Spinrad and Wing (1969)) should be treated with caution until the possible effects of saturation have been carefully considered. The enhancement of s-process elements is probably not confined to the Ba II and S stars; it occurs for example in o Vir (Williams, 1972), where it is unaccompanied by the enhancement of carbon characterising the classical Ba II and

CH stars. Variations in Na/Fe also occur and may be related to stellar evolution (Helfer and Wallerstein, 1968; P. M. Williams, 1971), although Cayrel de Strobel *et al.* (1970) have argued that this is a population effect due to enrichment of the interstellar medium. In M giants, there is an anomalously low abundance of water vapour which may (but need not) be due to a change in chemical composition connected with the CNO cycle, since the strength of H_2O at a given temperature depends on the ratio $(O-C)/H$ (Spinrad and Wing, 1969; Goon and Auman, 1970); the spectra of such cool stars are still only poorly understood, however. Another interesting aspect of stellar evolutionary effects is the distribution of lithium and beryllium (see Wallerstein and Conti, 1969).

A different kind of problem is posed by two groups of sharp-lined hot stars near the main sequence with strange atmospheric compositions, the peculiar (Ap) and metallic-lined (Am) stars. Ap stars (R. C. Cameron, 1967) include the magnetic spectrum variables, which are interpreted on the oblique rotator model as slowly rotating stars with some complicated distribution of magnetic fields and abundances over the surface; others show no detectable variation and there is a continuous gradation from normal to peculiar stars (Durrant, 1970). Typical peculiarities are low abundances of helium and oxygen and high ones of a great variety of elements including in various cases ^{3}He, Be, Si, P, Cr, Mn, Sr, rare earths, Pt, Hg and U with signs of anomalous isotopic composition of mercury (Dworetsky, 1970; Preston *et al.*, 1971). An identification of the unstable element promethium in the star HR 465 has been put forward (Aller and Cowley, 1970), but also challenged (Havnes and van den Heuvel, 1972; Aller and Cowley, 1972). Sargent and Searle (1967) have demonstrated that several abundance anomalies are closely related to atmospheric parameters, notably effective temperature. Am stars (Conti, 1970) are somewhat less dramatic than Ap stars, having characteristically a deficiency of Ca and Sc accompanied by over-abundance of iron and heavier elements. Most are members of close binary systems (perhaps because this is a means of ensuring slow rotation) and they seem to have no detectable magnetic field at their surfaces.

An obvious clue to the nature of Ap and Am stars is their exceptionally slow rotation combined with surface temperatures associated with shallow or non-existent surface convection zones, so that their visible abundance peculiarities are probably confined to a thin outer layer. Various nuclear explanations in the form of surface reactions, internal reactions followed by mixing or mass loss (Fowler *et al.*, 1965) and contamination by a companion that became a supernova (van den Heuvel 1967, 1968; Guthrie, 1972) have been tried, but run into serious difficulties. The most promising new idea in this field is that of element separation due to diffusion under the combined effects of gravity and radiation pressure, first suggested as a mechanism for depleting helium from the atmospheres of hot subdwarfs (Greenstein *et al.*, 1967) and afterwards investigated by Michaud (1970) as a general effect in Ap stars. Michaud showed that helium tends to sink down through the atmosphere because it has no lines or photo-ionisation continua at wavelengths where there is an appreciable outward radiative flux and that oxygen does the same at the low-temperature end of the Ap stars, where observation shows it to be deficient. Elements of sufficiently low intrinsic abundance can be leviated into the atmosphere by absorbing line radiation and it seems that photoionisation gives a good account of the behaviour of manganese and silicon.

The diffusion mechanism is promising in many ways, though it is not yet clear whether it can explain the whole range of abundance effects in Ap stars. For Am stars, however, a modification of the same idea provides a very satisfactory explanation (Watson, 1971; Smith, 1971). Here the diffusion does not occur in the atmosphere itself, which is rather turbulent, but in a stable radiative region underneath the hydrogen convection zone and above the HeII ionisation zone. This sandwich structure exists in more or less the right range of temperature to coincide with Am stars and has a suitable temperature for Mg, Ca and Sc to be ionised to inert-gas like configurations so that they sink while most rare elements are levitated by absorbing line radiation (Stickland and Whelan, 1972). A similar mechanism can account for the enhancement of ^{3}He in 3 Cen A (Michaud and Vauclair, 1972). Certain difficulties remain in understanding how the necessary degree of stability has been achieved, but the correspondence between element abundances and their electronic structure, together with the positions of Am stars in the HR diagram, make it very likely that this is the correct mechanism.

4. Population Effects in the Composition of Normal Stars

The term 'population effects' is used to denote effects on the atmospheric composition of normal stars that have not evolved to such an advanced stage that this is markedly affected by thermonuclear processes in the star itself; such effects are then usually attributed to the enrichment of the interstellar medium from which the stars were formed. The fundamental observational result here is the fact that, in our own neighbourhood, we see mostly (but not exclusively) stars of the disk population or Pop I for which the ages vary from 0 to about 10^{10} years, but the orbits round the centre of the Galaxy have small eccentricity and small inclination to the plane of the Milky Way and the chemical composition is not very different from that of the Sun with $-0.7 \leqslant [Fe/H] \leqslant +0.4$ approximately*. These stars are believed to have been born at various times from gas and dust spread out by its overall angular momentum into a disk shortly after the formation of our Galaxy by the collapse of an intergalactic cloud of gas. In addition we see in our neighbourhood a few stars with quite different kinematics and chemical composition which are all some 10^{10} yr old, have high orbital eccentricities and inclinations and a marked deficiency of heavy elements, $-3 < [Fe/H] \leqslant -0.6$. These are stars of the halo population or extreme Population II, also represented by the stars in globular clusters, which are believed to have been formed during the free-fall collapse while at the same time

---

* Square brackets are used to denote the logarithm (to base 10) of the ratio of some quantity in the atmosphere of the star to the same quantity in that of the Sun.

the interstellar medium was being enriched in heavy elements by explosive synthesis in supernovae and possibly other nuclear processes in short-lived, massive stars (Eggen et al., 1962). By looking at the composition of halo stars and also of disk stars with different ages, one hopes to be able to get some idea as to how the composition of the interstellar medium evolved as a function of time and thus to relate ideas on nucleosynthesis to ideas on the formation and evolution of stars and galaxies (Pagel, 1968, 1970).

Theoretical predictions as to how the composition of the medium should vary with time have been made on the basis of various assumptions by Fowler (1972), Truran and Cameron (1971) and Talbot and Arnett (1971); effects of mixing and spatial inhomogeneities are discussed by Reeves (1972). In Truran and Cameron's model, which is the most detailed, the bulk of heavy elements (C-Fe and r-process) is expected to increase with time (relative to hydrogen in the diminishing interstellar gas) at a fairly constant rate after an initial prompt enrichment, with more drastic variations with time in N and in elements formed by the s-process, but very little increase in He because the massive stars are assumed to evolve without mixing.

These predictions will now be compared with observation. The first question is that of the helium abundance, which is of special interest in connection with the possibility that most helium was synthesised in the Big Bang origin of the expanding universe (Wagoner et al., 1967). Excellent reviews of current knowledge of the distribution of helium have been given by Danziger (1970) and by Searle and Sargent (1972a). To summarise briefly, observations of helium lines in numerous young stars (Shipman and Strom, 1970; Norris, 1971; Leckrone, 1971; O'Mara and Simpson, 1972) give He/H very close to 0.10 or a mass fraction $Y=0.28$, with an error of perhaps 15%. Exactly the same value is found from radio and optical observations of recombination lines in HII regions (Churchwell and Mezger, 1970; Batchelor and Brocklehurst, 1972). The solar value of He/H is about 0.07, slightly less than 0.10, but the difference is only marginally significant. Ap and certain other not stars are deficient in helium in their atmospheres, while several other hot stars are exceptionally helium-rich; but this is probably due to special processes because the stars in question are often associated in clusters or binary systems with stars of normal helium abundance. Among the cooler stars, helium lines cannot be observed and the results from stellar-structure arguments are uncertain, so that there is little reliable information. Going now to the oldest stars -- those of extreme Population II – the situation is rather paradoxical. B subdwarfs near the extreme blue end of the horizontal branch of the Population II temperature-luminonsity diagram have weak helium lines indicating low atmospheric abundance, but detailed study reveals other peculairities reminiscent of Ap stars (Searle and Sargent, 1972a). Several arguments based on the theory of stellar structure imply $Y \simeq 0.3$ for practically all the globular clusters for which we have data and, incidentally, closely equal ages (Sandage, 1970). The main arguments for the high helium abundance are (i) the high-temperature boundary of the instability strip for RR Lyrae variables (Christy, 1968; van Albada and Baker, 1971); and (ii) relative numbers of stars on the red-giant and horizontal branches in the HR(temperature-luminosity) diagram (Iben and Rood, 1969, 1970). The position of the metal-deficient subdwarf Groombridge 1830, which is cool enough not to have evolved appreciably, below the Population I zero-age main sequence in the HR diagram also suggests that there is no significant reduction in helium content relative to Population I (Alexander and Pepper 1967; Cayrel 1968). We conclude that the initial helium content of Population II stars is high, and that weak-helium line stars are probably affected by diffusion. Further evidence for a pre-stellar origin of helium comes from the absence of significant variations in the He:H ratio deduced from observations of H$_{II}$ regions in external galaxies (Peimbert and Spinrad, 1970), even in the case of two blue compact galaxies in which both the stellar content and the abundances of oxygen and neon are exceptionally low (Searle and Sargent, 1972b). Lines due to helium in quasars of low red shift are weak, but this does not necessarily imply a low abundance (R. E. Williams, 1971). The distribution of helium in the universe is thus quite compatible with a cosmological origin along conventional lines.

The next question is that of a general enrichment of the interstellar medium in 'metals', i.e. the elements from carbon upwards. Following on from the discovery of weak spectral lines in Population II stars and the development of ideas on nucleo-synthesis in the 1950's, several attempts were made to correlate the composition of disk stars with their age deduced from the HR diagram or from their kinematic characteristics, which are also related to age, and, apart from a vague tendency of high-velocity stars to be mildly metal-deficient (Pagel, 1966), the results were rather inconclusive. More recently Spinrad and Taylor (1969) carried out narrow-band photometry of intense spectral features in red giants and identified a number of old 'super-metal-rich' (SMR) stars, both in and outside galactic clusters, that seemed to have a higher metal abundance than the relatively young Hyades cluster with [Fe/H]$\simeq$0.2, suggesting some anticorrelation between age and metal-deficiency. The results of Spinrad and Taylor have stirred up some controversy (Strom *et al.*, 1971; P. M. Williams, 1971; Blanc-Vaziaga *et al.*, 1973) and they may not give the correct [Fe/H] in all cases; but some old SMR stars certainly exist, occasionally with high space motions, e.g. 31 Aql (Hearnshaw,1971), and the issue is an important one mainly because apparently SMR stars seem to be an important component of the strong CN stellar population in the nuclear regions of certain galaxies (Spinrad *et al.*, 1971, 1972). The present situation can best be summarised by saying that some old stars are SMR, but we do not know how many. For stars very close to the Sun, they seem to be a small minority, since recent investigations of F and G stars by Powell (1972) and Hearnshaw 1972) respectively reveal that there is in fact a statistical correlation between age and metal deficiency among the disk stars (somewhat similar to that predicted by Truran and Cameron's model of 'prompt initial star formation'), though with a large and undoubtedly real scatter at any one age. We thus have a picture of continuing though non-uniform enrichment of the interstellar medium with time, such as may probably be understood on the basis of the considerations of Reeves (1972) which involve the mixing of expanding supernova remnants in certain active regions of star formation.

Given that the interstellar medium has been enriched, one may ask next whether abundances in normal stars give us any information as to details of the relevant nuclear processes. Models of explosive nucleosynthesis, which predict many of the Solar System relative abundances very well (Arnett, 1971), predict a correlation between odd-even ratios like Na/Mg in the products of synthesis and the overall mass of heavy elements, Z, initially present in the parent supernovae (Truran and Arnett, 1971). Likewise one might expect products of the s-process to be fed into the interstellar medium at a different rate from the C-Fe elements produced by explosive synthesis, since the s-process needs Fe-group nuclei to act as seeds and it may occur in stars having lower masses than supernovae and therefore evolving more slowly. We would expect, therefore, to find some correlation between the total amount of heavy elements present, represented by Fe/H, and the composition of the heavy element mixture itself represented by M_1/M_2 or M/Fe where M, M_1, M_2 represent some particular element.

Studies of the composition of Population I stars belonging to the galactic disk do indeed suggest the presence of odd-even effects well correlated with each other (Arnett, 1971), and partially correlated with Fe/H. The correlations may be perfectly real, but we have to remember that variations from the solar M_1/M_2 ratios are below a factor of 3 each way and our efforts to determine them (differentially relative to the Sun, since oscillator strengths are poorly known) are hag-ridden by the dependence of the solar and stellar curves of growth on microturbulence, collision damping and hyperfine structure. The extreme view is that of Unsöld (1969), who holds that all these differences in detailed composition are spurious. I don't share this view, but a good deal of work remains to be done before one can definitely say that M_1/M_2 differences of this order of magnitude are real.

In this connection it is instructive (and depressing) to look at various abundance determinations that have been made in extremely weak-lined stars of Population II, for which the parent interstellar medium was presumably so little enriched that we should expect the differential effects in M/Fe, say, to be the greatest (Table III). Differential effects among carbon and metals are small, and the contradictions between different analyses of the same star are often at least as big. Note also that Sr, Y and Zr, on the first magic-number peak of the s-process, are not collectively overdeficient by a significant amount relative to iron even in these extreme stars, although barium probably is. The supergiant BD + 39° 4926, which has been quoted as an example of extreme odd-even effects, has such a bizarre composition (slightly reminiscent of helium stars) that it is not a suitable specimen of what was happening in the interstellar medium, though it does seem to be an extremely interesting star in its own right. The conclusion, then, is that even for the most metal-deficient stars that we can see, that are usually assumed to be the oldest, we already encounter a substantial degree of mixing among elements attributed to quite different nuclear processes.

Is there any escape from this dreary pattern of a quasi-universal abundance distribution among the elements from carbon up? I believe there is, and that it is offered by nitrogen, which occupies a very special position in the chain of nucleo-

TABLE III

Specific abundance differences in some frequently analysed stars of very low metal abundance

| | Field Horizontal-branch 86986, 109995, 161817 (average) Kodaira 73 | Extreme sub-dwarfs | | | Yellow giant 122563 | | Wolffram 72 | He-C-N-O rich supergiant +39° 4926 Kodaira 73 |
		25329 Pagel-Powell 66	19445 Aller-Greenstein 60 Baschek 59, 63 Cohen-Strom 68	140283	Wallerstein et al. 63	Pagel 65; Ball and Pagel 67		
[Fe/H]	−1.7	−1.3	−1.9	−2.2	−2.7	−2.6	−2.7	−2.9
[C/Fe]	+0.7	−0.1		0.0		−1.0:	−0.4	+2.5
[Na/Fe]	+0.4:	−0.2			0.0	+0.5:	+0.3	+0.3
[Mg/Fe]	+0.8	+0.1	+0.9	+0.3	+0.2	0.0	0.0	+1.4
[Al/Fe]	−0.1		+0.1:	−0.0:	−0.1	+0.3	−0.1	−0.4
[Si/Fe]	+0.3		+0.6:	+0.1		+1.0:	+0.3:	+1.2
[Ca/Fe]	+0.7	+0.1	+0.3	+0.1	+0.1	0.0	+0.2	−0.7
[Sc/Fe]	+0.1	+0.4	0.0	+0.2	−0.6	−0.1	0.0	−0.1
[Ti/Fe]	+0.7	+0.1	+0.3	+0.2	−0.2	+0.1	+0.1	−0.4
[V/Fe]	−0.2:	−0.2	−0.2:	+0.2	−0.7	−0.6	−0.1	
[Cr/Fe]	+0.6	0.0	−0.1	+0.1	+0.2	0.0	0.0	+0.5
[Mn/Fe]	−0.2	−0.4	+0.2	+0.1	−0.6	−0.2	−0.2	
[Co/Fe]	+0.5:	0.0	+1.0:	+0.1:	−0.4	0.0	0.0	
[Ni/Fe]	0.0	−0.1	−0.1:	+0.3:	+0.1	+0.4	+0.2	
[Zn/Fe]					+0.7	+0.1	+0.3	
[Sr/Fe]	−0.2	−0.2	0.0	0.0	−1.2	−0.2	−0.6	−1.3
[Y/Fe]	+0.5				−0.6	0.0	−0.1	+0.1
[Zr/Fe]	−0.2				−1.3	−0.3	−0.1:	
[Ba/Fe]	+0.7	+0.4	−0.3	−0.6	−1.8	−1.0	−0.6	
[Ce/Fe]		+0.2:			−1.7	−1.1:		
[Eu/Fe]		0.0:			−1.7	−0.6:	−0.3:	

synthesis because it is not produced by explosive burning starting from H and He although it can be produced by ordinary H-burning from C to O if these are initially present (Howard et al., 1971). Furthermore, nitrogen has a very interesting behaviour among H II regions in external galaxies. Burbidge and Burbidge (1962, 1965) first noticed the great strength of [N II] relative to Hα in H II regions at the centres of several galaxies, and this was afterwards shown by Peimbert (1968) to be most probably a real abundance effect. More recently Searle (1971) has studied composition gradients among H II regions in the spiral arms of Sc galaxies and found large gradients in N/H, decreasing outwards from the centre, accompanied by a much smaller gradient in O/H which presumably represents heavy elements in general. So it is of interest to inquire whether the nitrogen abundance in stars shows a population effect distinct from that shown by other elements.

This question forms the goal of a study of CN band strengths in F-K type stars currently being carried out at the Royal Greenwich Observatory, using photoelectric measurements of the λ 4215 band made at Cambridge (Griffin and Redman, 1960) and David Dunlap Observatory (McClure and Van den Bergh, 1968; McClure, 1970) and photographic measurements of the stronger λ 3883 band observed at Herstmonceux, combined with model-atmosphere calculations by Fawell (1970) which predict the horizontal curve-of-growth shift [X] (equivalent to the logarithmic change in equivalent width of a weak line free from saturation effects) for a given set of atmospheric parameters(Harmer and Pagel, 1973). Figure 1 shows some results for Population I stars, in which the two photometric CN indices are plotted against [X] and compared with an empirical fiducial relation derived from standard stars with small [Fe/H] and

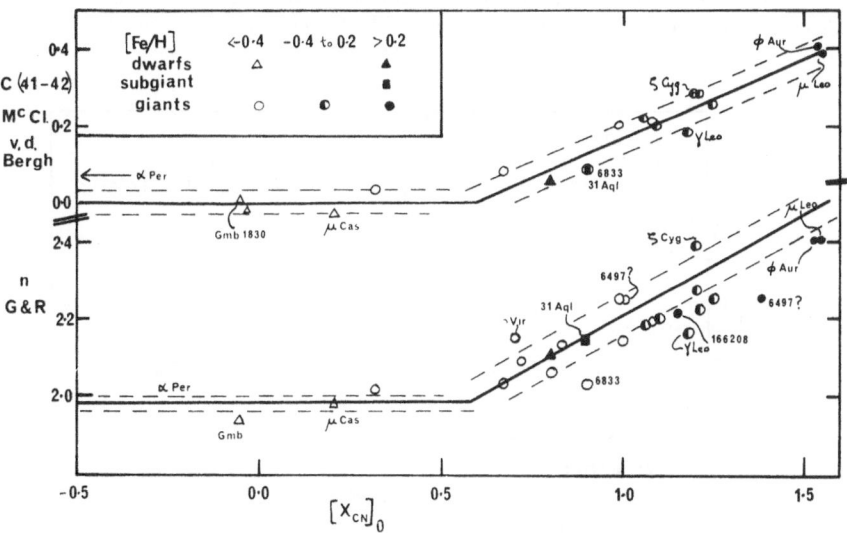

Fig. 1. Photoelectric λ 4215 CN band indices plotted against curve-of-growth shift [X] predicted on the 'null' hypothesis [C/Fe] = [N/Fe] = [O/Fe] = O with [Fe/H], T_{eff}, g known. Broken lines give a rough estimate of errors corresponding to ±0.1 in [X] on the sloping portion. The indices observed for α Persei are not due to CN.

well-determined atmospheric parameters. On the null hypothesis that C, N and O all behave like Fe in their abundance variation, each point should lie on the line, and one sees that, apart from a few interesting special cases, they do so with a deviation of only ± 0.15 or so in $\log X$. (However, μ Leo was plotted assuming that it was

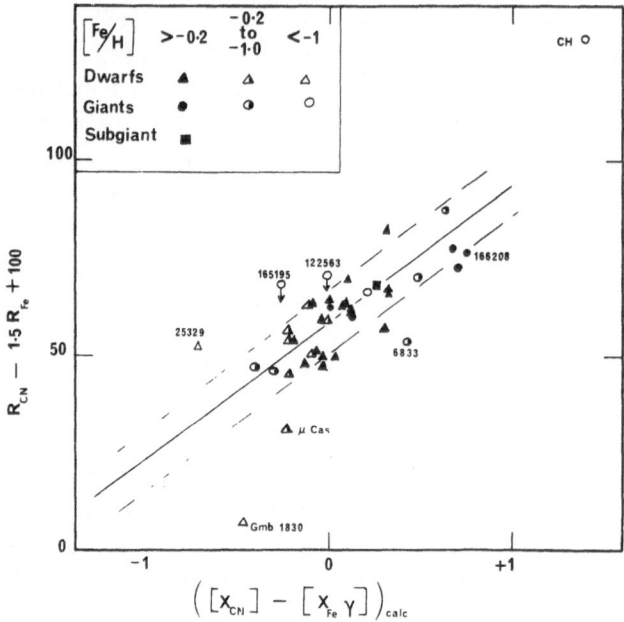

Figs. 2a–b. (a) Corrected photographic CN band dip measured at 60 A mm^{-1}, plotted against [X] defined as in Figure 1. Broken lines give a rough estimate of errors corresponding to ± 0.2 in [X]. (b) [N/Fe] plotted against [Fe/H] for stars measured by the photographic method. ν Indi (Harmer and Pagel, 1970) has been added.

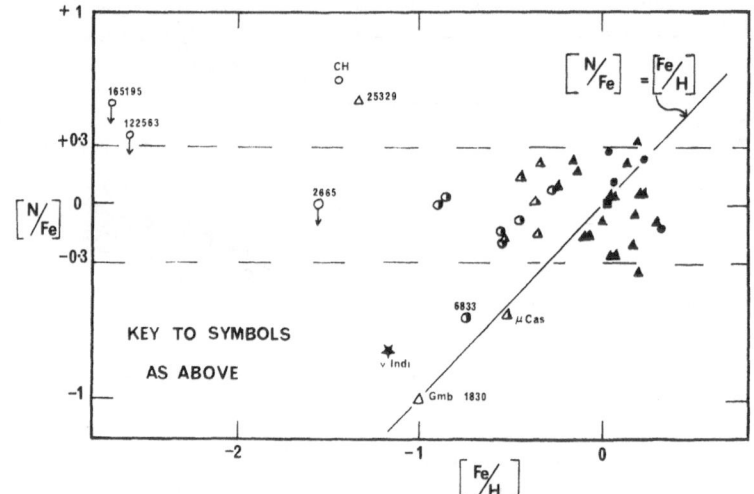

Fig. 2b. Key to symbols as in Figure 2a.

SMR; so if it is not SMR, then it is presumably abnormally N-rich.) Thus no marked changes in N/Fe are evident in the stars predicted to have strong CN bands, although Fe/H varies by about an order of magnitude. However, the stars that we might expect to be expecially interesting, namely weak-lined stars of Population II, have such weak CN bands in any case that this particular criterion cannot be applied and we have to use our photographic measurements of the 3883 band, which we compare with two strong iron lines nearby in order to remove uncertainties in the continuum level and in [Fe/H] so as to derive [N/Fe] (Harmer, 1972).

The results of doing this are shown in Figure 2a and 2b, in which the upper panel shows the strength of CN corrected for the strength of the iron features with, again, an empirical calibration line based on the null hypothesis [N/Fe]=0. For stars with halo kinematics, especially when Fe/H is very low, marked deviations from the line now appear, though in the one case of HD 25329 this deviation is in the wrong direction (i.e. high N/Fe). This is a perfectly real effect (cf. Pagel, 1970), but it is also quite unique because all other weak-lined stars (apart from CH stars) have been found to have very weak CN bands, so it should not deflect us from perceiving a general trend for Population II stars to fall below the line. Another halo star with low N-abundance is ν Indi (Harmer and Pagel, 1970; Bell, 1970). The lower panel gives a plot of [N/Fe] against [Fe/H] for all the stars studied by the photographic method. Stars with disk kinematics, even when [Fe/H] is quite low, show no special effect, but the halo stars (discounting those that are peculiar or strongly affected by blends) show a distinct tendency to lie along the line [N/Fe]=[Fe/H]. (Tomkin, 1972 has derived a similar result for Groombridge 1830 on the basis of a high-dispersion analysis).

These effects are large enough to be offered as a challenge to the universal-abundance viewpoint and as an indication that, in some respects, one can actually see effects of differential thermonuclear enrichment of the interstellar medium in the atmospheric compositions of normal unevolved stars.

References

GENERAL REFERENCES

Alexander, J. B. and Pepper, S.: 1967, *Observatory* **87**, 267.
Aller, L. H.: 1972, *Ann. New York Acad. Sci.* **194**, 45
Aller, M. F. and Cowley, C. R.: 1970, *Astrophys. J.* **163**, L45
Aller, M. F. and Cowley, C. R.: 1972, *Astron. Astrophys.* **19**, 286.
Aller, L. H. and Greenstein, J. L.: 1960, *Astrophys. J. Suppl.* **5**, 139.
Arnett, W. D.: 1971, *Astrophys. J.* **166**, 153.
Auer, L. H. and Mihalas, D.: 1972, *Astrophys. J. Suppl.* **24**, 193, No. 205.
Ball, C. and Pagel, B. E. J.: 1967, *Observatory* **87**, 19.
Baschek, B.: 1959, *Z. Astrophys.* **48**, 95.
Baschek, B.: 1963, *Z. Astrophys.* **56**, 207.
Batchelor, A. S. J. and Brocklehurst, M.: 1972, *Astrophys. Letters* **11**, 129.
Bell, R. A.: 1970, *Monthly Notices Roy. Astron. Soc.* **150**, 15.
Bell, R. A. and Upson, W. L., II.: 1971, *Astrophys. Letters* **9**, 109.
Blanc-Vaziaga, M. J., Cayrel, G., and Cayrel, R.: 1973, *Astrophys. J.* **180**, 871.
Boesgaard, A. M.: 1970, *Astrophys. J.* **161**, 163.

Boyer, R., Henoux, J. C., and Sotirovsky, P.: 1971, *Solar Phys.* **19**, 330.

Bridges, J. M. and Wiese, W. L.: 1970, *Astrophys. J.* **161**, L71.

Burbidge, E. M. and Burbidge, G. R.: 1957, *Astrophys. J.* **126**, 357.

Burbidge, E. M. and Burbidge, G. R.: 1962, *Astrophys. J.* **135**, 694.

Burbidge, E. M. and Burbidge, G. R.: 1965, *Astrophys. J.* **142**, 634.

Burbidge, E. M., Burbidge, G. R., Fowler, W. A., and Hoyle, F.: 1957, *Rev. Mod. Phys.* **29**, 547.

Cameron, A. G. W.: 1968, in L. H. Ahrens (ed.), *Origin and Distribution of the Elements*, Pergamon Press, New York, p. 125.

Cameron, R. C. (ed.): 1967, *The Magnetic and Related Stars*, Mono Book Corp., Baltimore.

Cayrel, R.: 1968, *Astrophys. J.* **151**, 997.

Cayrel de Strobel, G., Chauve-Godard, J., Hernandez, G., and Vaziaga, M. J.: 1970, *Astron. Astrophys.* **7**, 408

Christy, R. F.: 1968, *Quart. J. Roy. Astron. Soc.* **9**, 13.

Churchwell, E. and Mezger, P. G.: 1970, *Astrophys. Letters* **5**, 227.

Cohen, J. G. and Strom, S. E.: 1968, *Astrophys. J.* **151**, 623.

Conti, P. S.: 1970, *Publ. Astron. Soc. Pacific* **82**, 781.

Crawford, H. J., Price, P. B., and Sullivan, J. D.: 1972, *Astrophys. J.* **175**, L149.

Danziger, I. J.: 1970, *Ann. Rev. Astron. Astrophys.* **8**, 161.

De Boer, K. S., Olthoff, H., and Pottasch, S. R.: 1972, *Astron. Astrophys.* **16**, 417.

Dufton, P. L.: 1972, *Astron. Astrophys.* **16**, 301.

Dupree, A.: 1972, *Astrophys. J.* **178**, 527.

Durrant, C. J.: 1970, *Monthly Notices Roy. Astron. Soc.* **147**, 75.

Dworetsky, M.: 1970, *Bull. Amer. Astron. Soc.* **2**, 311.

Eggen, O. J., Lynden-Bell, D., and Sandage, A. R.: 1962, *Astrophys. J.* **136**, 748.

Fawell, D. R.: 1970, *Thesis*, University of Sussex.

Fichtel, C. E.: 1971, *Phil. Trans. Roy. Soc. London* **270**, 167.

Fowler, W. A.: 1972, in F. Reines (ed.), *Cosmology, Fusion and Other Matters: Gamow Memorial Volume*, Hilger, London, p. 67.

Fowler, W. A., Burbidge, E. M., Burbidge, G. R., and Hoyle, R.: 1965, *Astrophys. J.* **142**, 423.

Foy, R.: 1972, *Astron. Astrophys.* **18**, 26.

Fujita, Y.: 1970, *Interpretation of Spectra and Atmospheric Structure in Cool Stars*, Tokyo University Press and University Park Press, Baltimore and Manchester.

Garz, T. and Kock, M.: 1969, *Astron. Astrophys.* **2**, 274.

Garz, T., Kock, M., Richter, J., Baschek, B., Holweger, H., and Unsöld, A.: 1969, *Nature* **223**, 1254.

Gasson, R. E. M. and Pagel, B. E. J.: 1966, *Observatory* **86**, 196.

Gingerich, O., Noyes, R. W., Kalkofen, W., and Cuny, Y.: 1971, *Solar Phys.* **18**, 347.

Goldberg, L., Müller, E. A., and Aller, L. H.: 1960, *Astrophys. J. Suppl.* **5**, 1.

Goon, G. and Auman, J. R.: 1970, *Astrophys. J.* **75**, 785.

Greene, T. F.: 1969, *Astrophys. J.* **157**, 737.

Greene, T. F.: 1970, *Astrophys. J.* **161**, 365.

Greene, T. F., Perry, J., Snow, T. P., and Wallerstein, G.: 1973, *Astron. Astrophys.* **22**, 293.

Greenstein, G. S., Truran, J. W., and Cameron, A. G. W.: 1967, *Nature* **213**, 871.

Griffin, R. F. and Redman, R. O.: 1960, *Monthly Notices Roy. Astron. Soc.* **120**, 287.

Guthrie, B. N. G.: 1972, *Astrophys. Space Sci.* **15**, 214.

Hall, D. N. B., Noyes, R. W., and Ayres, T. R.: 1972, *Astrophys. J.* **171**, 615

Hardorp, J. and Scholz, M.: 1970, *Astrophys. J. Suppl.* **19**, 193 (no. 173).

Harmer, D. L.: 1972, M. Sc. Dissertation, Sussex University.

Harmer, D. L. and Pagel, B. E. J.: 1970, *Nature* **255**, 349.

Harmer, D. L. and Pagel, B. E. J.: 1973, *Monthly Notices Roy. Astron. Soc.*, in press.

Hauge, Ö.: 1972, *Solar Phys.* **26**, 263.

Havnes, O. and Van den Heuvel, E. P. J.: 1972, *Astron. Astrophys.* **19**, 283.

Hearnshaw, J.: 1971, *Astrophys. J.* **168**, 109.

Hearnshaw, J.: 1972, *Mem. Roy. Astron. Soc.* **77**, 55.

Helfer, H. L. and Wallerstein, G.: 1968, *Astrophys. J. Suppl.* **16**, 1.

Herbig, G. H.: 1970, in L. Houziaux and H. E. Butler (eds.), 'Ultra-Violet Stellar Spectra and Related Ground-Based Observations', *IAU Symp.* **36**, 315.

Hirayama, T.: 1971, *Solar Phys.* **19**, 384.
Howard, W. M., Arnett, W. D., and Clayton, D. D.: 1971, *Astrophys. J.* **165**, 495.
Iben, I., Jr.: 1965, *Astrophys. J.* **142**, 1447.
Iben, I., Jr. and Rood, R. T.: 1969, *Nature* **223**, 933.
Iben, I., Jr. and Rood, R. T.: 1970, *Astrophys. J.* **161**, 587.
Johnson, H. L. and Méndez, M.E.:1970, *Astron. J.* **75**, 785.
Kodaira, K.: 1973, *Astron. Astrophys.* **22**, 273.
Kodaira, K. S. and Scholz, M.: 1970, *Astron. Astrophys.* **6**, 93.
Lambert, D. L., Mallia, E. A., and Petford, D.: 1971 *Monthly Notices Roy. Astron. Soc.* **154**, 265.
Leckrone, D. S.: 1971, *Astron. Astrophys.* **11**, 387.
McClure, R. D.: 1970, *Astron. J.* **75**, 41.
McClure, R. D. and Van den Berg, S.: 1968, *Astron. J.* **73**, 313.
Merrill, P. W.: 1952, *Astrophys. J.* **116**, 21.
Michaud, G.: 1970, *Astrophys. J.* **160**, 641.
Michaud, G. and Vauchlair, S.: 1972, *Astrophys. Letters* **11**, 117.
Mihalas, D.: 1970, *Stellar Atmospheres*, W. H. Freeman, San Franscisco.
Mihalas, D.: 1972, *Astrophys. J.* **176**, 139.
Mogro-Compero, A. and Simpson, J. A.: 1972, *Astrophys. J.* **171**, L5.
Norris, J.: 1971, *Astrophys. J. Suppl.*, no. 197.
O'Mara, B. J. and Simpson, R. W.: 1972, *Astron Astrophys.* **19**, 167.
Osterbrock, D.: 1970, *Quart. J. Roy. Astron. Soc.* **11**, 199.
Pagel, B. E. J.: 1965, *Roy. Obs. Bull.* No. 104.
Pagel, B. E. J.: 1966, in H. Hubenet (ed.), 'Abundance Determination in Stellar Spectra', *IAU Symp.* **26**, 359, Academic Press.
Pagel, B. E. J.: 1968, in L. H. Ahrens (ed.), *Origin and Distribution of the Elements*, Pergamon Press, p. 195.
Pagel, B. E. J.: 1970, *Vistas in Astronomy* **12**, 313; *Quart. J. Roy. Astron. Soc.* **11**, 172.
Pagel, B. E. J.: 1971a, in D. Mihalas, B. E. J. Pagel, and P. Souffrin (eds.), *Theorie des Atmospheres Stellaires*, Geneva Observatory, p. 180.
Pagel, B. E. J.: 1971b, *J. Phys. B* **4**, 279.
Pagel, B. E. J. and Powell, A. L. T.: 1966, *R. Obs. Bull.*, No. 124.
Parsons, S. B.: 1967, *Astrophys. J.* **150**, 263.
Peery, B. F.: 1971, *Astrophys. J.* **163**, L1.
Peimbert, M.: 1968, *Astrophys. J.* **154**, 33.
Peimbert, M. and Costero, R.: 1969, *Bol. Obs. Tonanzintla y Tacubaya* **5**, 3.
Peimbert, M. and Spinrad, H.: 1970, *Astrophys. J.*, **159**, 809.
Peters, G. J. and Aler, L .H.: 1970, *Astrophys. J.* **159**, 525.
Pottasch, S. R. 1968, in L. H. Ahrens (ed.), *Origin and Distribution of the Elements*, Pergamon Press, p. 183.
Powell, A. L. T. 1972, *Monthly Notices Roy. Astron. Soc.* **155**, 483.
Preston, G. W., Vaughan, A. H., White and R. E., Swings, J. P.: 1971, *Pub. Astron Soc. Pacific* **83**, 607.
Querci, M. and Querci, F.: 1970, *Astron. Astrophys.* **9**, 1.
Reeves, H.: 1972, *Astron. Astrophys.* **19**, 215.
Sandage, A. R.: 1970, *Astrophys. J.* **162**, 841.
Sargent, W. L. W. and Searle, L.: 1967, in R. C. Cameron (ed.), *The Magnetic and Related Stars*, p. 209.
Searle, L.: 1971, *Astrophys. J.* **168**, 327.
Searle, L. and Sargent, W. L. W.: 1972a, *Comments Astrophys. Space Phys.* **4**, 59.
Searle, L. and Sargent, W. L. W.: 1972b, *Astrophys. J.* **173**, 25.
Shipman, H. L. and Strom, S. E.: 1970, *Astrophys. J.* **159**, 183.
Smith, M.: 1971, *Astron. Astrophys.* **11**, 325.
Spinrad, H. and Taylor, B. J.: 1969, *Astrophys. J.* **157**, 1279.
Spinrad, H. and Wing, R. F.: 1969, *Ann. Rev. Astron Astrophys.* **7**, 249.
Spinrad, H., Gunn, J. E., Taylor, B. J., McClure, R. D., and Young, J. W.: 1971, *Astrophys. J.* **164**, 11.
Spinrad, H., Smith, H. E., and Taylor, D. J.: 1972, *Astrophys. J.* **175**, 649.
Stickland, D. J. and Whelan, J. A. J.: 1972, *Monthly Notices Roy. Astron. Soc.* **155**, 11P.
Strom, S. E., Strom, K. M., and Carbon, D. F.: 1971, *Astron. Astrophys.* **12**, 177.

Takens, R. J.: 1970, *Astron. Astrophys.* **5**, 244.
Talbot, R. J., Jr. and Arnett, W. D.: 1971, *Astrophys. J.* **170**, 409.
Thompson, R. I., Schnopper, H. W., and Rose, W. K.: 1971, *Astrophys. J.* **163**, 533.
Tomkin, J.: 1972, *Monthly Notices Roy. Astron. Soc.* **156**, 349.
Truran, J. W. and Arnett, W. D.: 1971, *Astrophys. Space Sci.* **11**, 430.
Truran, J. W. and Cameron, A. G. W.: 1971, *Astrophys. Space Sci.* **14**, 179.
Unsöld, A.: 1942, *Z. Astrophys.* **21**, 22.
Unsöld, A.: 1969, *Science* **163**, 1015.
Urey, H. C.: 1967, *Quart. J. Roy. Astron. Soc.* **8**, 23.
Van Albada, T. S. and Baker, N.: 1971, *Astrophys. J.* **169**, 311.
Van den Heuvel, E. P. J.: 1967, 1968, *Bull. Astron. Inst. Neth.* **19**, 11, 309, 326, 432, 449.
Van Paradijs, J.: 1972, *Nature Phys. Sci.* **238**, 37.
Wagoner, R. V., Fowler, W. A., and Hoyle, F.: 1967, *Astrophys. J.* **148**, 3.
Wallerstein, G.: 1969, *Astrophys. J.* **158**, 607.
Wallerstein, G. and Conti, P. S.: 1969, *Ann. Rev. Astron. Astrophys.* **7**, 99.
Wallerstein, G. and Greenstein, J. L.: 1964, *Astrophys. J.* **139**, 700.
Wallerstein, G., Greenstein, J. L., Parker, R., Helfer, L. H., and Aller, L. H.: 1963, *Astrophys. J.*
    137, 280.
Warner, B.: 1965, *Monthly Notices Roy. Astron. Soc.* **129**, 263.
Warner, B.: 1967, *Monthly Notices Roy. Astron. Soc.* **137**, 119.
Watson, W. D.: 1971, *Astron. Astrophys.* **13**, 263.
Whaling, W., King, R. B., and Martinez-Garcia, M.: 1969, *Astrophys. J.* **158**, 389.
Wiese, W. L., Smith, M. W., and Glennon B. M.: 1966, *Atomic Transition Probabilities, Vol. I; Hydro-
    gen Through Neon*, NBS, Washington D.C.
Williams, P. M.: 1971, *Monthly Notices Roy. Astron. Soc.* **155**, 215.
Williams, P. M.: 1972, *Monthly Notices Roy. Astron. Soc.* **155**, 17P.
Williams, R. E.: 1971, *Astrophys. J.* **167**, L27.
Wolffram, W.: 1972, *Astron. Astrophys.* **17**, 17.
Wolnick, S. J., Berthel, R. O., and Wares, G. W.: 1970, *Astrophys. J.* **162**, 1037.
Wolnick, S. J., Berthel, R. O., and Wares, G. W.: 1971, *Astrophys. J.* **166**, L31.
Worrall, G. and Wilson, A. M.: 1972, *Nature*, **236**, 15.
Yamashita, Y.: 1972, *Publ. Astron. Soc. Japan*, **24**, 49.

REFERENCES TO TABLE I

[1] Goldberg, L., Muller, E. A., and Aller, L. H.: 1960, *Astrophys. J. Suppl.* **5**, 1.
[2] Bertsch, D. L., Fichtel, C. E., and Reames, D. V.: 1972, *Astrophys. J.* **171**, 169. Numbers have
    been normalised to give a rough overall best fit to photospheric values.
[3] de Boer, K. S., Olthof, H., and Pottasch, S. R.: 1972, *Astron. Astrophys.* **16**, 417.
[4] Dupree, A. K.: 1972, *Astrophys. J.* **178**, 527.
[5] Cameron, A. G. W.: 1968, in L. H. Ahrens (ed.), *Origin and Distribution of the Elements*,
    Pergamon Press, p. 125. Numbers have been normalised to give a rough overall best fit to
    photospheric values for Na, Mg, Al, Si, Ca.
[6] Grevesse, N.: 1968, *Solar Phys.* **5**, 159.
[7] Engvold, O., Kjeldseth Moe, O., and Maltby, P.: 1970, *Astron. Astrophys.* **9**, 79.
    Stellmacher, G., and Wiehr, E.: 1971, *Solar Phys.* **21**, 97.
[8] Hauge, O. and Engvold, O.: 1968, *Astrophys. Letters* **2**, 235.
[9] Lambert, D. L.: 1968, *Monthly Notices Roy. Astron. Soc.* **138**, 143.
[10] Lambert, D. L. and Swings, J. P.: 1967, *Solar Phys.* **2**, 34.
[11] Andrews, M. and Mugglestone, D.: 1963, *Monthly Notices Roy. Astron. Soc.* **125**, 347.
    Baschek, B. and Holweger, H.: 1967, *Z. Astrophys.* **67**, 143.
[12] Mugglestone, D. and O'Mara, B. J.: 1965, *Monthly Notices Roy. Astron. Soc.* **129**, 41.
[13] Mallia, E. A.: 1968, *Solar Phys.* **3**, 505.
[14] Müller, E. A., Baschek, B., and Holweger, H.: 1968, *Solar Phys.* **3**, 125.
[15] Lambert, D. L. and Warner, B.: 1968, *Monthly Notices Roy. Astron. Soc.* **138**, 181.
[16] Holweger, H.: 1971, *Astron. Astrophys.* **10**, 128.
[17] Lambert, D. L. and Warner, B.: 1968, *Monthly Notices Roy. Astron. Soc.* **140**, 197.
[18] Lambert, D. L. and Warner, B.: 1968, *Monthly Notices Roy. Astron. Soc.* **139**, 35.

[19] Grevesse, N. and Swings, J. P.: 1972, *Astrophys. J.* **171**, 179.
[20] Swings, J. P., Lambert, D. L., and Grevesse, N.: 1969, *Solar Phys.* **6**, 3.
[21] Lambert, D. L., Mallia, E. A., and Brault, J.: 1971, *Solar Phys.* **19**, 289.
[22] Holweger, H.: 1972, *Solar Phys.* **25**, 14.
[23] Lambert, D. L. and Mallia, E. A.: 1969, *Solar Phys.* **10**, 311.
[24] Warner, B.: 1968, *Monthly Notices Roy. Astron. Soc.* **138**, 229.
     Warner, B.: 1972, *Observatory* **92**, 50.
[25] Cocke, C. L., Curnutte, B., and Brand, J. H.: 1971, *Astron. Astrophys.* **15**, 299.
[26] Blackwell, D. E., Collins, B. S., and Petford, D.: 1972, *Solar Phys.* **23**, 292.
[27] Garz, T., Holweger, H., Kock, M., and Richter, J.: 1969, *Astron. Astrophys.* **2**, 446.
     Baschek, B., Garz, R., Holweger, H., and Richter, J.: 1970, *Astron. Astrophys.* **4**, 229.
[28] Nussbaumer, H. and Swings, J. P.: 1970, *Astron. Astrophys.* **7**, 455.
[29] Foy, R.: 1972, *Astron. Astrophys.* **18**, 26.
[30] Garz, T.: 1971, *Astron. Astrophys.* **10**, 175.
[31] Grevesse, N. and Swings, J. P.: 1970, *Solar Phys.* **13**, 19.
[32] Kock, M. and Richter, J.: 1968, *Z. Astrophys.* **69**, 180.
[33] Lambert, D. L., Mallia, E. A., and Warner, B.: 1969, *Monthly Notices Roy. Astron. Soc.* **142**, 71.
[34] Ross, J. E. and Aller, L. H.: 1970, *Proc. Natl. Acad. Sci. U.S.A.* **66**, 983.
[35] Lambert, D. L. and Mallia, E. A. 1968, *Monthly Notices Roy. Astron. Soc.* **140**, 13.
[36] Hauge, O.: 1972, *Solar Phys.* **26**, 273, 276.
[37] Ross, J. E., and Aller, L. H.: 1972, *Solar Phys.* **25**, 30.
[38] Grevesse, N. and Blanquet, G.: 1969, *Solar Phys.* **8**, 5 (comprehensive discussion of solar rare-earth abundances based on oscillator strengths by Corliss and Bozman which are probably untrustworthy). Tb and Ho have not been definitely identified in the solar spectrum and the results for Er and Lu (each based on one line only) are omitted from Table I.
[39] Molnar, H.: 1972, *Astron. Astrophys.* **20**, 69. (Oscillator strengths from Corliss and Bozman.)
[40] Ross, J. and Aller, L. H.: 1972, *Solar Phys.* **23**, 13. (The result is based on a single ultra-violet line of doubtful oscillator strength.)
[41] Grevesse, N.: 1970, *Geochim. Cosmochim Acta* **34**, 1131.
[42] Grevesse, N.: 1969, *Solar Phys.* **6**, 381.

THEORIES OF NUCLEOSYNTHESIS

J. W. TRURAN*

Belfer Graduate School of Science, Yeshiva University, New York, N.Y., U.S.A.

Abstract. A review is presented of current theories of nucleosynthesis. The predicted contributions from (1) cosmological nucleosynthesis, (2) super-massive stars, (3) non-violent (quasi-static) stellar evolution, (4) supernova explosions, (5) cosmic ray interactions with the interstellar medium and (6) nova explosions to the observed solar system abundances are summarized. Recent studies of 'explosive nucleosynthesis' in supernovae and of the production of lithium, beryllium and boron by the inter-action of cosmic rays with interstellar gas are emphasized. Observations of stellar spectra which either impose limitations upon or provide confirmation of various aspects of these theories are noted, as are several critical nuclear experiments. The general picture which emerges is incouraging in that most of the major abundance features appear to be at least qualitatively understood, but significant further research is required.

1. Introduction

Theories of nucleosynthesis have been guided, historically, by the prevailing knowl-edge of element abundances. The earliest spectral studies of the Sun and stars were able to establish only that the elements of which they and the Earth are composed are qualitatively the same. It was therefore quite reasonable to assume a universe of uniform chemical composition and to search within the framework of cosmology for a set of physical conditions which would account for the present abundance distribution (see for example the review article by Alpher and Herman, 1950). Difficulties associated with 'bridging' the gaps at masses $A=5$ and 8 (no stable nuclear species exist with these mass numbers) soon made it clear that universal 'big-bang' synthesis (Alpher *et al.* 1948) could not have been responsible for the formation of the bulk of the nuclear species heavier than helium ($A=4$). The current status of the big-bang theory of nucleosynthesis, emphasizing its promising role in the production of ^2D, ^3He, ^4He and ^7Li, is reviewed in Section 2.

The essential role played by thermonuclear processes in providing an energy source sufficient to account for stellar lifetimes of billions of years was established in the late 1930s by the calculation of Bethe (1939) and von Weizsacker (1938). The fusion of four protons to one helium nucleus ('hydrogen burning') taking place in 10% of the Sun's mass releases several orders of magnitude more energy than is available as gravitational potential energy. It was not immediately recognized, however, that subsequent thermonuclear burning phases in stellar interiors might account for the formation of many of the heavier nuclei observed in nature. The recognition that nucleosynthesis is a continuing process in stellar interiors followed the discovery by Merrill (1952) of the presence of the element technetium in the atmospheres of red giant stars. As technetium has no stable isotopes (the longest lived isotope, ^{98}Tc, having a half life $\tau=1.5 \times 10^6$ y), its presence confirms that thermonuclear processes involving heavy nuclei have very recently taken place in the interior.

---

* Present address: Department of Astronomy, University of Illinois, Urbana, Illinois.

A. G. W. Cameron (ed.), Cosmochemistry, 23–49. All Rights Reserved

This conclusion has been confirmed by many subsequent spectroscopic analyses which have revealed that pronounced abundance variations occur in certain broad classifications of stars. 'Carbon stars', for example, showing anomalous atmospheric abundances of $^{13}C(^{12}C/^{13}C\sim4)$ and nitrogen (N/C$\sim$35) compared to the solar values, are interpreted as stars in which hydrogen burning by means of the CNO cycles has taken place in the interior with the resulting materials being mixed to the surface by convection.

The broader classifications of stars in our Galaxy into disc (Population I) and halo (Population II) populations is reinforced by considerations of their relative metal contents Z (Z being the fraction by mass in the form of all elements heavier than ^{4}He). Disc population stars typically show heavy element concentrations within a factor three of the Sun ($Z_{\odot}=0.015$), while halo stars may have somewhat lower metal contents and a few $extreme$ halo population stars are known to have metal contents $Z\sim10^{-3}-10^{-2}\ Z_{\odot}$. As the halo stars are generally assumed, for dynamic reasons, to constitute an earlier galactic generation, these abundance variations suggest that the heavy elements observed in solar system matter represent the integrated effects of stellar or supernova nucleosynthesis over the lifetime of the Galaxy. The fact that relatively few stars are observed with extremely low metal contents ($Z<10^{-1}\ Z_{\odot}$) suggests that the net rate of heavy element synthesis during the earliest phases of galactic history may have been somewhat greater than at present and certainly that relatively few stars of lifetimes greater than ten billion years (masses $\lesssim$ one solar mass ($M_{\odot}$)) were formed at this epoch (Schmidt, 1963; Truran and Cameron, 1971). The possible role of nucleosynthesis in supermassive stars in the earliest stages of galactic history is briefly discussed in Section 3, and the contributions predicted as a consequence of the non-violent (quasi-static) evolution of normal stars throughout galactic history are reviewed in Section 4.

Recent theoretical calculations regarding two specific mechanisms – supernova nucleosynthesis and cosmic-ray-induced nucleosynthesis – are elaborated in Sections 5 and 6, respectively. As will become clear, the final catastrophic supernova phase of evolution characteristic of stars in certain mass ranges seems the most promising site for the synthesis of the bulk of the heavy elements observed in nature. The interaction of high energy protons and α-particles in the cosmic radiation with the more abundant heavy constituents of the interstellar medium (carbon, nitrogen, oxygen and neon) has recently been demonstrated to produce significant abundances of the isotopes of lithium, beryllium and boron – nuclei which had previously presented severe difficulties. Finally, in Section 7 a brief discussion of the possible role of nova explosions in nucleosynthesis is presented.

The major boundary conditions with which we are provided for theories of nucleosynthesis are shown in Figure 1 – the solar system abundances (Cameron, 1968). It may be useful at this stage to scan the important abundance features and to keep these in mind in our subsequent discussions. The two most abundant nuclei, hydrogen and ^{4}He, together with their isotopes deuterium and ^{3}He, seems most likely to have emerged from the cosmological big-bang (Section 2). Lithium, beryllium and

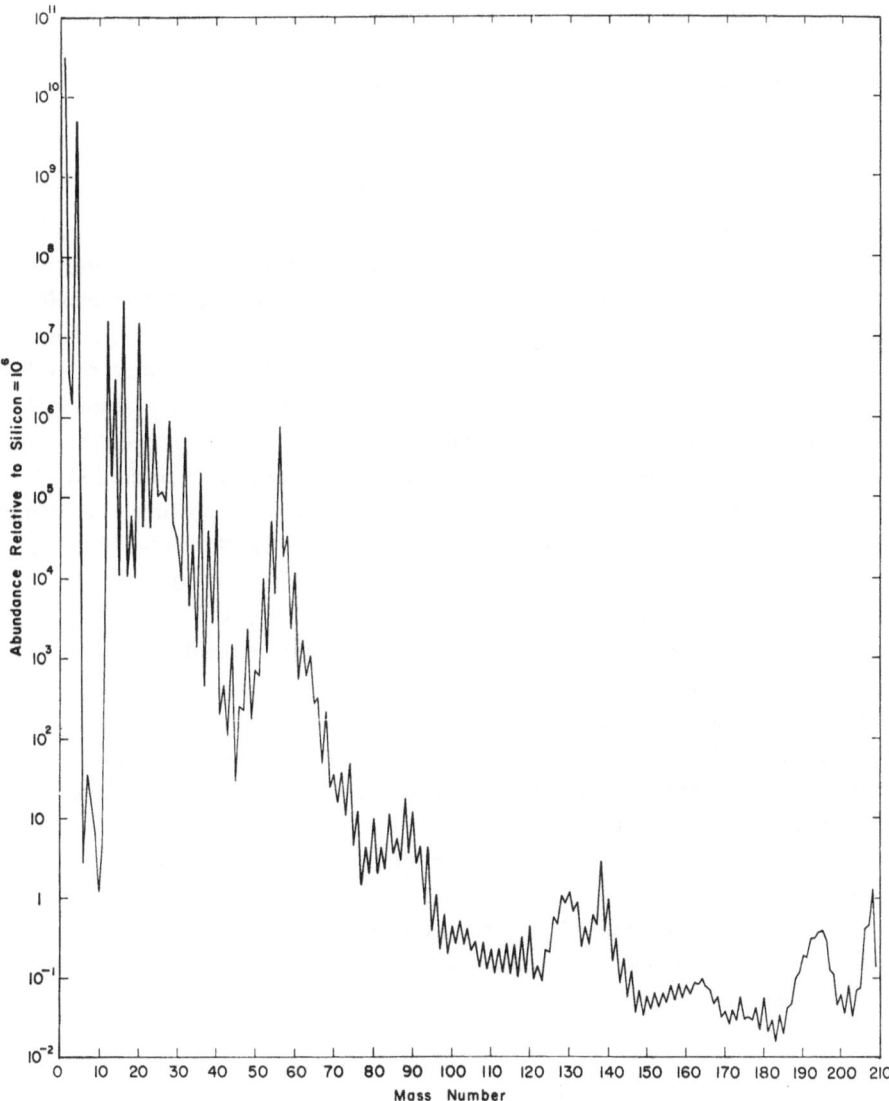

Fig. 1.   The solar system abundances (Cameron, 1968) are plotted as a function of mass number.

boron, long a challenge to nucleosynthesis theories even though extremely low in abundance compared to neighboring elements, have recently been explained as resulting from the interaction of cosmic rays with the interstellar medium (6). The large abundances of ^{12}C and ^{16}O are interpreted as the products of stellar helium burning. The subsequent explosion of these fuels in supernovae (Section 5) at temperatures in the range $T \sim 2 - 8 \times 10^9$ K can account for most of the observed abundance features in the mass region $20 \lesssim A \lesssim 60$, including the 'equilibrium' peak centered on iron. The formation of most nuclei past mass 60 is generally attributed to neutron-capture processes. We shall see that the abundance features at mass numbers $A \sim 88$, 130, 138, 195 and 208 which stand out above an otherwise rather constant abundance level may be very naturally interpreted in terms of neutron shell structure and its implied nuclear stability.

2. Cosmological Nucleosynthesis

Interest in the subject of the universal synthesis of the elements was revived by the discovery of Penzias and Wilson (1965) of background micro-wave radiation and its interpretation as evidence for a universal background temperature of approximately 3 K. The predictions of big-bang nucleosynthesis have since been re-examined by Peebles (1966a, b), Wagoner et al. (1967), Greenstein (1968) and most recently by Wagoner (1973). One very firm conclusion follows from all these studies: for a present day temperature of ~ 3 K and a present day universal density $\lesssim 10^{-28}$ g cm^{-3} (this upper limit being inferred from cosmological red-shift observations), the universal synthesis of the heavy elements in the big bang is not possible. The production of heavier elements would require either a significantly higher present day density or a lower present day temperature. The triple-alpha reaction $(3\,^4\mathrm{He} \rightarrow {}^{12}\mathrm{C})$, which serves successfully to bridge the mass gaps at $A=5$ and 8 in stellar helium burning, is ineffective at the densities achieved in the expanding fireball at the appropriate burning temperature.

Significant concentrations of ^{4}He and other light elements *can* be formed under appropriate conditions in the primordial fireball. The abundances of hydrogen, deuterium, ^{3}He, ^{4}He and ^{7}Li resulting from element production in low density universes with zero leptonic number, as calculated by Wagoner et al. (1967), are summarized in Table I. Here the mass fractions of ^{1}H and ^{4}He together with the ratios $(^2\mathrm{D}/^1\mathrm{H})$, $(^3\mathrm{He}/^4\mathrm{He})$ and $(^7\mathrm{Li}/^1\mathrm{H})$ relative to their current solar system values* are tabulated as a function of present day density for a present day temperature $T_0 = 2.7$ K. Significant helium production occurs for all densities $\gtrsim 10^{-32}$ g cm^{-3} (this is generally not the case for degenerate-neutrino universes).

TABLE I
Cosmological nucleosynthesis

Identification	Density ($T_0 = 2.7$ K)						
	3×10^{-33}	10^{-32}	3×10^{-32}	10^{-31}	3×10^{-31}	10^{-30}	3×10^{-30}
^{1}H	0.95	0.89	0.81	0.76	0.75	0.74	0.73
^{4}He	0.032	0.098	0.19	0.24	0.25	0.26	0.27
$(^2\mathrm{D}/^1\mathrm{H})/(^2\mathrm{D}/^1\mathrm{H})_{\odot}$	530	340	130	27	4.0	0.54	0.012
$(^3\mathrm{He}/^4\mathrm{He})/(^3\mathrm{He}/^4\mathrm{He})_{\odot}$	110	32	8.4	2.4	1.4	0.72	0.32
$(^7\mathrm{Li}/^1\mathrm{H})/(^7\mathrm{Li}/^1\mathrm{H})_{\odot}$	0.063	0.44	0.50	0.080	0.033	0.70	3.4

* The Cameron (1968) values are used for ^{1}H, ^{4}He and ^{7}Li. The ^{2}D and ^{3}He abundances are those arrived at by Geiss and Reeves (1972) from an analysis of solar and solar wind abundances: $^2\mathrm{D}/^1\mathrm{H} = 3 \times 10^{-6}$ and $^3\mathrm{He}/^4\mathrm{He} = 10^{-4}$.

We note that there exists a particularly interesting universe $(\varrho \sim 3 \times 10^{-31} \text{ g cm}^{-3})$ for which reasonable agreement with solar system abundances is also achieved for ^{2}D and ^{3}He. The ^{7}Li abundances predicted by more recent calculations (Wagoner, 1972) are somewhat higher, but universal synthesis still fails by a factor $\lesssim 10$ to produce sufficient ^{7}Li for this interesting case.

3. Nucleosynthesis in Supermassive Stars

Observational evidence that the oldest stars and clusters in our Galaxy have heavy element abundances ranging from less than one percent to a substantial fraction of the solar value strongly suggests that heavy element synthesis must have taken place either very early in the galaxy's history or even perhaps in pregalactic events (Truran and Cameron, 1971). Cosmological big-band nucleosynthesis, as we have seen, cannot produce even the lowest observed abundances $(Z \sim 10^{-3} Z_\odot)$. As an alternative to a pregalactic event, Fowler and Hoyle (1964), Wagoner et al. (1967), and most recently Wagoner (1968, 1969) have explored the nucleosynthesis conditions which might be expected to occur in supermassive stars, $M \gtrsim 10^3 M_\odot$. Their calculations indicate that ^{4}He can be produced in substantial amounts in supermassive stars if the peak temperatures reach $T = 10^{10}$ K. Heavy elements may also be formed in sufficient amounts to explain the very low levels $(\lesssim 10^{-2} Z_\odot)$ observed in the oldest Population II stars, but the relative isotopic and elemental abundances may be quite different. Furthermore, in ultra-fast stellar explosions (the ejection velocities considerably exceed escape velocities) comparable abundance levels $(\sim 10^{-2} Z_\odot)$ of nuclei from ^{12}C to ^{28}Si may be achieved with relative abundances which are consistent with solar system data (Wagoner, 1968, 1969).

4. Nucleosynthesis in Stellar Evolution

The evolution of a star is defined by a sequence of alternate stages of gravitational contraction to higher temperatures and thermonuclear burning of the available fuel at these temperatures. The various thermonuclear processes in stars which can contribute, in successive burning stages, to the production of the heavier elements observed in nature were first defined by Burbidge et al. (1957) and by Cameron (1957). In what follows we will define each of these processes and briefly summarize the current theoretical situation.

4.1. Hydrogen Burning

The greatest part of the active burning lifetime of a star $(\gtrsim 90\%)$ is spent converting hydrogen to helium in a core comprising roughly 10% of its mass at temperatures ranging from 10 to 50 million degrees. Two complex nuclear reaction sequences can participate in this burning – the proton-proton chains and the carbon-nitrogen-oxygen (CNO) bi-cycle in which these heavier nuclei act as catalysts in the conversion of hydrogen to helium. The proton-proton reactions dominate the burning in stars of mass less than or comparable to the Sun, while the CNO-cycles come into play in more massive stars which typically burn hydrogen at higher temperatures.

Interest in the proton-proton burning reactions has recently been stimulated by the results of the solar neutrino detection experiment of Davis (1955). The reaction sequences illustrated in Table II indicate that, following competitive branches involv-

TABLE II

Proton-proton chains

PP I
$$^1H + {}^1H \rightarrow {}^2D + e^+ + \nu$$
$$^2D + {}^1H \rightarrow {}^3He + \gamma$$
$$^3He + {}^3He \rightarrow {}^4He + 2{}^1H + \gamma$$
⟨ or

PP II
$$^3He + {}^4He \rightarrow {}^7Be + \gamma$$
$$^7Be + e^- \rightarrow {}^7Li + \nu$$
$$^7Li + {}^1H \rightarrow {}^8Be^* \rightarrow 2{}^4He$$
or

PP III
$$^7Be + {}^1H \rightarrow {}^8B + \gamma$$
$$^8B \rightarrow {}^8Be^* + e^+ + \nu$$
$$^8Be^* \rightarrow 2{}^4He$$

ing ^3He and ^7Be, the positron decay of ^8B yields a neutrino of maximum energy 14.06 MeV. In the Davis experiment, impinging neutrinos from this reaction are expected to interact with the ^{37}Cl contained in an enormous tank of C_2Cl_4 (cleaning fluid) to form ^{37}Ar by the reaction $^{37}C. + \nu \rightarrow e^- + {}^{37}Ar$. The ^{37}Ar atoms thus formed can be extracted efficiently and counted. The results to date (Davis, 1972; Davis et al., 1971) provide only an upper limit on the counting rate of 1 SNU (1 SNU $= 10^{-36}$ captures per second per target atom) but this alone represents a severe discrepancy with respect to theory. Existing models for the sun which incorporate the most recent nuclear cross section data, realistic opacities and our best estimates of the helium and heavy element contents of the Sun (see for example Abraham and Iben (1971) and Bahcall and Ulrich (1971)) predict counting rates of 9.9 SNU and 9.0 SNU, respectively. with uncertainties due to parameter changes of the order of ± 5 SNU. A recent revision of the rate of the p-p reaction (Gari and Huffman, 1972) reduces the flux predicted by the calculations of Bahcall and Ulrich to 7.1 SNU.

This discrepancy poses a very severe challenge to nuclear astrophysics. The input nuclear and opacity data for solar models have been carefully scrutinized, and variations within what are *believed* to be realistic limits of uncertainty to not resolve this problem. The neutrino flux is also dependent upon the heavy-element content of the sun but, as pointed out by Bahcall and Ulrich (1971), it would be necessary to assume a primordial value $\lesssim 10^{-3}$ (by mass), compared to the observed surface value of 1.5×10^{-2}, to reduce the predicted counting rate to within experimental limits. The inability to readily explain this discrepancy by more conventional means has led to speculations concerning more fundamental aspects of solar and neutrino physics. Fowler (1972) has noted, for example, that periodic variations in the sun's central temperature on time scales $\lesssim 3 \times 10^7$ yr (the time required for photons to diffuse from the center to the surface), triggered by convective instabilities, would be consistent with a reduced central temperature for the present sun. As both the $^3\text{He}(^4\text{He},\gamma)^7\text{Be}$ and the $^7\text{Be}(p,\gamma)^8\text{B}$ branching reactions are temperature sensitive, a 10% reduction in the present central temperature which determines the neutrino flux would largely resolve the current discrepancy with experiment. Further theoretical calculations together with a re-examination of the nuclear and opacity data are clearly required before any definitive statements can be made.

An inportant consequence of hydrogen burning by the CNO bi-cycle (Table III) is the synthesis of ^{14}N. Detailed calculations of CNO-burning (Caughlan, 1965) reveal that generally ^{14}N will be the most abundant nucleus under equilibrium burning conditions; this is due to the relatively low rate of the $^{14}\text{N}(p,\gamma)^{15}\text{O}$ reaction. Most of the initial carbon and oxygen present in the hydrogen-burning region will thus be converted to ^{14}N by this mechanism. This is extremely important for nucleosynthesis, as the formation of ^{12}C and ^{16}O by means of helium burning reactions is not accompanied by the production of amounts of ^{14}N consistent with the solar system abundances. The formation of ^{14}N therefore represents a 'secondary' process of nucleosynthesis, requiring ^{12}C and ^{16}O nuclei formed in a previous generation of stars to have been mixed into the interstellar gas from which the CNO-burning star has since formed.

There is rather convincing observational evidence for CNO-cycle burning in stars. Wallerstein et al. (1967) have analyzed the star HD 30353, which shows a very high abundance of nitrogen and a deficiency of hydrogen. They interpret the nitrogen as ^{14}N which has been formed in CNO-burning in the interior and mixed to the surface of the star by convection. They have determined the nitrogen-to-oxygen abundance ratio in this star to be approximately 50. This is in marked contrast to the ratio $\text{N}/\text{O} \simeq 0.1$ characteristic of solar system material, but is quite consistent with the ratio expected ($\text{N}/\text{O} \simeq 60$) for CNO-burning under typical hydrogen-burning conditions in stars (Caughlan, 1965). Similarly, the ratio of ^{12}C to ^{13}C observed in carbon stars (Climenhaga, 1958: Wyler, 1965) is about 3 or 4 (as compared with roughly 90 for solar system abundances), which is almost exactly the inverse ratio of the interaction probabilities for the $^{12}\text{C}(p,\gamma)^{13}\text{N}$ and $^{13}\text{C}(p,\gamma)^{14}\text{N}$ reactions.

4.2. HELIUM BURNING

The helium burning phase of evolution of a star is dominated by two reactions – the 'triple-alpha' reaction ($3\,^4\text{He} \rightarrow\,^{12}\text{C}$) which forms ^{12}C and the α-particle capture reaction $^{12}\text{C}(\alpha,\gamma)^{16}\text{O}$ which destroys it. The subsequent evolution of a star is critically dependent upon the relative abundances of ^{12}C and ^{16}O in the core following helium exhaustion and these abundances, in turn, are sensitive to the relative rates of these two reactions. The rate of the triple-alpha reaction (Salpeter, 1952) has recently been reduced somewhat by a new experimental determination of the energy of the second excited state of ^{12}C (Austin *et al.*, 1971; McCaslin *et al.*, 1972) through which the reaction proceeds, and is now assumed to be known to acceptable accuracy.

Of greater concern is the large uncertainty associated with the rate of carbon destruction. Under typical astrophysical conditions ($T \sim 10^8$ K, $\varrho \sim 10^4$ g cm^{-3}), the capture of an α-particle by ^{12}C proceeds primarily in the tail of the 7.12 MeV excited state of ^{16}O. Recent experimental determinations of the reduced alpha width of this state span the range ~ 0.02 to 0.8 of the single particle limit (Puhlhofer *et al.*, 1970; Jaszczak and Macklin, 1970; Jaszczak *et al.*, 1970; Weisser *et al.*, 1971), defining the magnitude of uncertainty of the reaction cross section. Within these limits, the abundances predicted to result from helium burning vary dramatically, ranging effectively from pure ^{12}C (for the choice ~ 0.02) to pure ^{16}O (Deinzer and Salpeter, 1964; Vidal *et al.*, 1971).

Until this problem is resolved by further nuclear experimentation, the details of stellar evolution past the helium burning phase remain highly uncertain. For the purposes of our subsequent discussion we assume the correct values lies in the vicinity of ~ 0.1, giving rise to roughly comparable amounts of ^{12}C and ^{16}O following helium burning.

4.3. CARBON AND OXYGEN BURNING

Burbidge *et al.* originally proposed that helium burning would be followed by an 'α-process' in which α-particles released by photodisintegrations at temperatures $T \gtrsim 10^9$ K would be recaptured past ^{20}Ne to form heavier nuclei by the sequence of reactions ^{20}Ne$(\alpha,\gamma)^{24}$Mg, ^{24}Mg$(\alpha,\gamma)^{28}$Si, etc. We now know that before this can take place, nuclear burning proceeds by the interaction of carbon and oxygen nuclei with themselves ('heavy-ion' reactions). At a temperature $T_9 \sim 0.8$ (T_9 is the temperature in units 10^9 K), ^{12}C interacts with itself by the reactions

$$^{12}\text{C} + {}^{12}\text{C} \rightarrow \begin{cases} {}^{20}\text{Ne} + \alpha \\ {}^{23}\text{Na} + p \\ {}^{23}\text{Mg} + n \end{cases}$$

giving roughly comparable yields of alphas and protons with a weak ($\sim 5\%$) endothermic neutron branching. The cross-section has been studied experimentally (Patterson *et al.*, 1969; Mazarakis and Stephens, 1972) but uncertainties of a factor ~ 5 remain in extrapolations to lower energies (Michaud and Vogt, 1972 Michaud, 1972). At slightly higher temperatures, $T_9 \lesssim 2$, oxygen nuclei can interact with one another as

$$^{16}\text{O} + {}^{16}\text{O} \rightarrow \begin{cases} {}^{28}\text{Si} + \alpha \\ {}^{31}\text{P} + p \\ {}^{31}\text{S} + n \end{cases}$$

A recent experimental study of the cross sections (Spinka and Winkler, 1972) shows the proton and alpha yields to be roughly comparable with a weak neutron branching ($< 10\%$). We note that the $^{12}C + ^{16}O$ reaction can also contribute in these burning stages (Stephens and Mazarakis, 1970; Hansen and Zaidins, 1971; Michaud, 1972).

Evolutionary models for carbon-burning and oxygen-burning stars have been calculated by numerous authors. While the role of these burning stages in the hydrostatic evolution of more massive stars is established, serious questions have arisen concerning the extent of their contributions to nucleosynthesis. Detailed calculations both of carbon-burning (Arnett and Truran, 1969) and of oxygen-burning (Woosley et al., 1972) nucleosynthesis indicate that the resulting distributions of primary products, lying in the mass range $20 \lesssim A \lesssim 40$, are not in accord with solar system abundances. In contrast, calculations of carbon- and oxygen-burning nucleosynthesis under explosive conditions predict abundance distributions which agree remarkably well with solar system data (Arnett, 1969; Truran and Arnett, 1970). A discussion of these explosive burning calculations is presented in Section 5.

4.4. SILICON BURNING

The primary products of hydrostatic carbon and oxygen burning are the self-conjugate (α-particle) nuclei ^{20}Ne, ^{24}Mg, ^{28}Si and ^{32}S. As core contraction increases the central temperature ($T_9 \gtrsim 2$), readjustments due to photodisintegration and capture reactions first favor an increased concentration of ^{28}Si. At still higher temperatures ($T_9 \sim 3$) the burning of ^{28}Si proceeds, not by the interaction of ^{28}Si with itself (the coulomb barrier is prohibitively large) but rather by a complex sequence of nuclear reactions. Photodisintegration of ^{28}Si and lighter nuclei release neutrons, protons and α-particles which can be recaptured on the remaining ^{28}Si and heavier nuclei, gradually shifting the abundance peak toward the position of iron. These readjustments tend toward the minimum energy configuration – a 'nuclear statistical equilibrium' – which, at the temperatures and densities of interest ($T_9 \sim 3 - 5$, $\varrho \sim 10^5 - 10^8$ g cm^{-3}) corresponds to a peak in the vicinity of iron (^{56}Fe is the nucleus possessing the maximum binding energy per nucleon). The role of such an equilibrium (seeking) process in the formation of the iron-peak nuclei was first emphasized by Hoyle (1946). Burbidge et al. (1957) adopted this viewpoint and explored physical conditions providing the best fits to the abundances of the iron group nuclei. A typical equilibrium fit to the observed iron-peak abundances is shown in Figure 2. Detailed calculations of the thermonuclear reaction sequences involved in silicon burning and the accompanying approach to nuclear statistical equilibrium have since been performed by several authors (Truran et al., 1966; Bodansky et al., 1968; Michaud and Fowler, 1972).

While hydrostatic silicon burning may serve as a stable energy generation phase of evolution in more massive stars (Rakavy et al., 1967), it is very difficult to see how the iron-peak nuclei thus formed can be removed from the star and contribute to nucleosynthesis. As nuclei in this region are characterized by the highest binding energies per nucleon, no nuclear energy is available from further burning. The core contraction which must follow should result in severe distortion of these abundance patterns. The ultimate fate of these stars is uncertain, but it seems clear that the equilib-

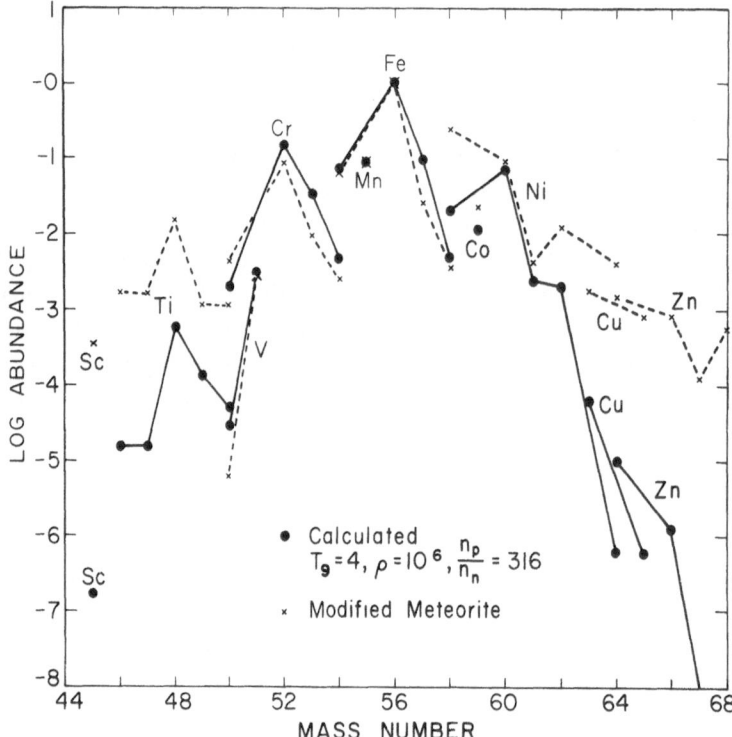

Fig. 2. A typical equilibrium calculation of the abundances of iron-peak nuclei (Cameron, 1963) is compared to solar system abundances. This equilibrium configuration is specified by a temperature $T = 4 \times 10^9$ K, density $\varrho = 10^6$ g cm^{-3}, and ratio of free protons to free neutrons $n_p/n_n = 316$. The designation 'modified meteorite' implies that the iron abundance has been reduced by a factor of 5 to agree with the solar photospheric value.

rium peak formed in stable burning will be destroyed in the subsequent evolution. We will demonstrate in the next section that the formation of the iron-peak nuclei observed in nature under explosive burning conditions is more likely, for under these conditions synthesis and ejection from the star can occur simultaneously.

4.5. HEAVY ELEMENT SYNTHESIS

Three mechanisms are considered to be primarily responsible for the formation of the heavy elements ($A > 60$). The s-process (neutron capture on a *slow* time scale) is characterized by mean times between successive neutron captures ($\sim 10^5$ y) which are long compared to typical beta-decay lifetimes near the valley of beta stability. Intervening beta decays therefore constrain the buildup to the vicinity of the valley of beta stability, and the s-process capture flow path proceeds as illustrated in Figure 3 for a rather typical region of the nuclear chart. No production of nuclei past mass $A = 210$ results from this neutron-capture process. The r-process (neutron capture on a *rapid* or fast time scale) is defined by capture time scales which are short compared to beta-decay lifetimes. Successive neutron captures in this instance can build off the

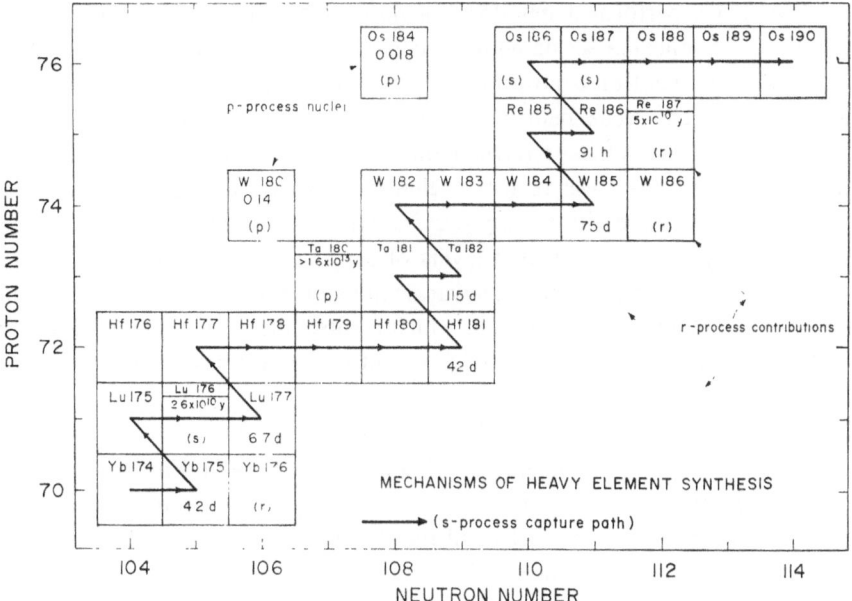

Fig. 3. A representative section of the nuclear chart in the heavy element region. The s-process capture path is illustrated, and the neutron-deficient and neutron-rich stable nuclei which can be formed only by p-process or r-process synthesis, respectively, are identified.

valley of beta stability into the neutron rich region. Following exhaustion of the neutron flux, the capture products approach the position of the valley of beta stability by beta decay; such nuclei as ^{176}Yb, ^{186}W and ^{187}Re which lie off the s-process capture path can be produced by this mechanism. We note finally that the nuclei ^{180}Ta, ^{180}W and ^{184}Os which lie on the neutron-deficient side of the valley of beta stability have been 'bypassed' by both neutron capture processes. These nuclei are typically low in abundance compared to neighboring neutron-capture products. In principle they can be produced by photodisintegration processes or proton capture processes (p-process) operating on the products of previous neutron capture.

Both the r-process, which requires an enormous neutron flux, and the p-process, which requires higher temperatures, are generally assumed to occur in supernova events; these processes will be discussed further in Section 5. Neutron sources sufficient to give rise to the required s-process neutron flux levels can arise during earlier phases of evolution. A relatively weak neutron flux is provided by the ^{13}C$(\alpha,n)^{16}$O reaction occurring during core helium burning in stars which previously burned hydrogen via the p-p chains and have retained their small primordial ^{13}C abundances. In more massive stars hydrogen burning takes place via the CNO-cycles resulting in the conversion of the bulk of the initial CNO nuclei to ^{14}N. A far more substantial neutron source then becomes available during helium burning by the reactions

$$^{14}N(\alpha,\gamma)^{18}F(e^+\nu)^{18}O(\alpha,\gamma)^{22}Ne(\alpha,n)^{25}Mg.$$

Peters (1968) has performed detailed calculations of s-process nucleosynthesis in stars of 9 and 15 solar masses based upon contributions from this source, concluding that the maximum neutron exposures that can be achieved are less than that of solar system material. The weak neutron branches from carbon and oxygen heavy ion interactions may also contribute (Arnett and Truran, 1969), but detailed integrations of the neutron exposures over realistic stellar models have not yet been performed.

A far more promising site for s-process synthesis is provided by the convective mixing of the hydrogen and helium layers of a star in the wake of helium-burning shell flashes in red-giant envelopes (Schwarzschild and Harm, 1967). The admixture of protons into a helium burning zone in which a substantial ^{12}C abundance has been formed can result in the production of free neutrons by the reactions

$$^{12}C(p,\gamma)^{13}N(e^+ v)^{13}C(\alpha,n)^{16}O$$

as described in detail by Sanders (1967) and Cameron and Fowler (1971). This mechanism has the additional advantage that the s-process products thus formed should be readily mixed to the surface of the star where they may be lost as part of a less violent stellar mass loss process; they are not asked to survive the extreme shock wave heating which one expects to be associated with ejection in a supernova event.

Rather convincing evidence exists for s-process nucleosynthesis in stars. The presence of technetium lines in the spectra of red giants (Merrill, 1952) can best be explained in terms of recent neutron capture synthesis. The Ba II and CH stars are identified by excesses of heavy elements, particularly barium and the rare earths, compared to iron (Danziger, 1965; Wallerstein and Greenstein, 1964). The variations in abundance of the mercury isotopes relative to solar observed in several Hg stars (Dworetsky et al., 1970; Preston, 1971) also seem to demand that neutron capture processes have been active.

Predicted correlations between neutron-capture cross sections and s-process abundances (Gibbons and Macklin, 1968; Allen et al., 1972) also provide a test of the s-process hypothesis. In the 'local approximation' for the s-process (Clayton et al., 1961), one predicts a constancy of the products $\sigma_{n,\gamma}N_s$, where N_s is the abundance of the s-process nucleus and $\sigma_{n,\gamma}$ is its neutron capture cross section. This behavior has now been confirmed, building upon measurements of the required capture cross sections at energies $\sim$30–100 keV (Macklin and Gibbons, 1965; Allen et al., 1972). The peaks in the solar system abundance pattern at masses $A \sim$ 88, 138 and 208 (attributable to closed neutron shells at $N = 50$, 82 and 126) follow naturally from this picture, as cross sections for capture out of these regions are extremely small.

4.6. Synthesis of Lithium, Beryllium and Boron

The process or processes (then unknown) which are responsible for the synthesis of deuterium, lithium, beryllium and boron were designated collectively the 'x-process' by Burbidge et al. (1959). The challenge was to find some mechanism capable of compensating for the destruction of these elements, and to a lesser extent ^{3}He, by proton interactions at temperatures below those typical of hydrogen burning.

Deuterium, for example, is readily destroyed at temperatures $T \gtrsim 10^6$ K by the reaction $^2D(p,\gamma)^3He$. Any initial concentrations of ^6Li, ^7Li, ^9Be, ^{10}B and ^{11}B will also be readily destroyed by (p,α) reactions in the contraction phase at temperatures below 10^7 K. We have noted previously that cosmological nucleosynthesis provides a likely mechanism for the production of ^2D, ^3He and perhaps some ^7Li. In Section 6, recent calculations of the production of ^6Li, ^7Li, ^9Be, ^{10}B and ^{11}B by means of cosmic-ray interactions will be described in some detail.

Several other possible mechanisms deserve mention. Spallation processes in stellar atmospheres may contribute to the production of some of these light nuclei, but quantitative estimates of the extent of these contributions are difficult. ^3He production may take place as a consequence of shell hydrogen burning, particularly if material in these regions can be mixed to the surface by convection (as in the Schwarzschild and Harm models). Substantial production of ^7Li may also result from these shell-flashing configurations (Cameron and Fowler 1971) if ^3He can be mixed into the helium layers, by the reactions $^3He(\alpha,\gamma)^7Be(e^-,\nu)^7Li$.

5. Supernova Nucleosynthesis

While the supernova phase constitutes only an extremely small fraction of the lifetimes of stars in restricted mass ranges, it nevertheless appears that a substantial fraction of the abundances of heavy elements ($A \gtrsim 20$) observed in nature have resulted from such events. Hydrodynamic studies of possible supernova mechanisms predict very promising conditions both for the synthesis of elements through the vicinity of iron by charged-particle reactions and for the neutron-capture synthesis of heavier nuclei. The energy requirements for the synthesis of heavier nuclei (nuclear trans-formation past iron represent net endoergic processes) are easily consistent with observed energies associated with supernova explosions ($\sim 10^{51}$ to 10^{52} ergs, or the equivalent of the total energy output of the sun burning at its present rate over a lifetime in excess of 12 b.y.). From the point of view of nucleosynthesis, one has the additional advantage that nuclear transformations proceed in the shock wave ejection of the core and envelope material, assuring that the resulting abundance distributions will not be distorted by subsequent evolution. In contrast, it is very diffcult to under-stand how iron-peak nuclei formed in core silicon burning might be expelled from a star in their original form.

Three mechanisms of nucleosynthesis generally demand temperature and/or density conditions which are most likely to be achieved in supernova explosions – the charge-particle synthesis of nuclei from neon to iron, the **r**-process and the **p**-process. Recent calculations pertaining to these three mechanisms will be reviewed in the following discussion.

5.1. EXPLOSIVE CHARGED-PARTICLE NUCLEOSYNTHESIS

Recent calculations of carbon-, oxygen- and silicon-burning nucleosynthesis under explosive conditions have proved successful in their predictions of detailed abundance

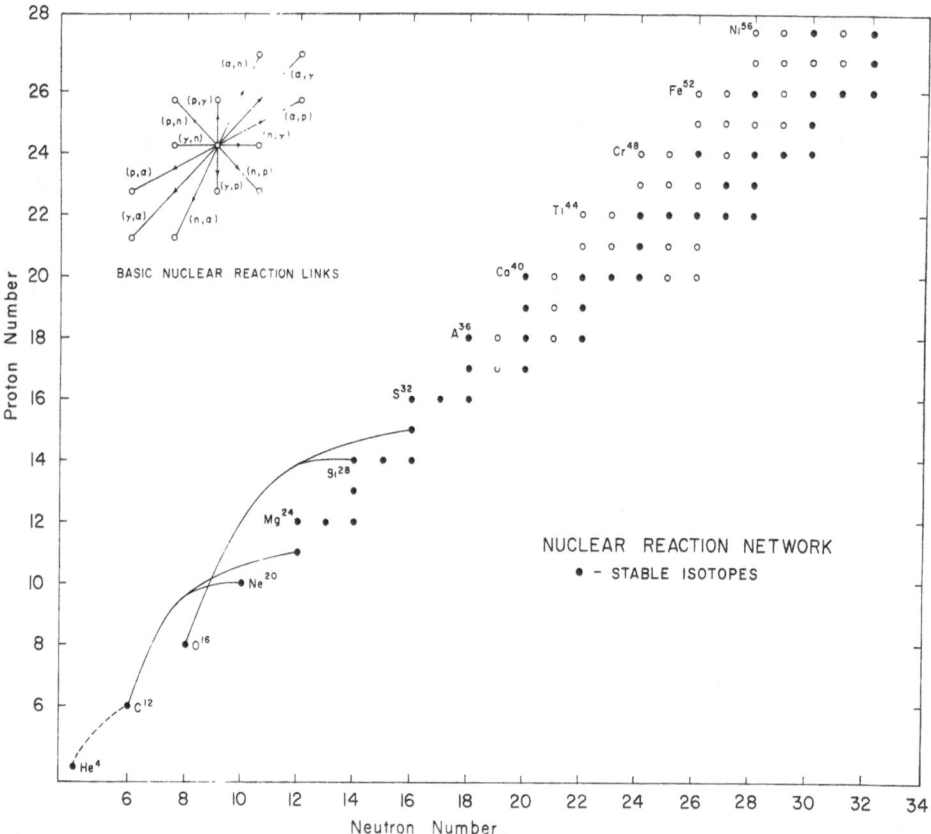

Fig. 4. Nuclear reaction network used in the investigation of explosive oxygen-burning nucleo-synthesis (Truran and Arnett, 1970). In addition to the triple-alpha, $^{12}C + ^{12}C$ and $^{16}O + ^{16}O$ reactions, all reactions involving neutrons, protons, α-particles and photons were included in this study.

features. In these studies one chooses an initial composition consistent with post helium-burning conditions in a stellar core and then explores the range of temperature and density conditions consistent with the expulsion of the matter in a supernova-type event for those conditions which provide the best fits to the solar system abundances. The expansion time scale is equated with the hydrodynamic time scale $\tau \sim 446/\varrho^{1/2}$ s. Complex nuclear reaction networks such as that illustrated in Figure 4 are required to follow in detail the nuclear reactions which occur. For an initial composition consisting of equal parts by mass ^{12}C and ^{16}O with a small admixture of ^{18}O * the following results are obtained: for temperatures $T_9 \sim 2 (\varrho \sim 10^5$ g cm^{-3}), explosive 'carbon' burning will reproduce very well many of the observed abundance features in the

* A mass fraction ~ 0.02 of ^{18}O is consistent with the following prior history of this core material: (1) all initial CNO-nuclei (present in amounts consistent with solar system matter) were converted to ^{14}N during hydrogen burning and (2) the ^{14}N was destroyed during helium burning by $^{14}N(\alpha, \gamma)^{18}F(e^+\nu)^{18}O$.

mass range $20 \leqslant A \leqslant 30$ (Arnett, 1969a; see also Hansen, 1971); at slightly higher temperatures ($T_9 \sim 3.6$, $\varrho \sim 5 \times 10^5$ g cm^{-3}), explosive 'oxygen' burning will reproduce most of the observed features in the mass range $28 \leqslant A \lesssim 44$ (Truran and Arnett, 1970); for temperatures $T_9 > 4.5$ K ($\varrho \gtrsim 10^6$ g cm^{-3}) the iron-peak region $48 \lesssim A \lesssim 60$ may be formed in an explosive 'silicon' burning event (Truran et al., 1967; Truran and Arnett, 1971).

The abundance distributions resulting from these explosive burning stages are compared to the solar system abundances (Cameron, 1968) in Figures 5, 6 and 7. Explosive carbon burning gives rise to a distribution of abundances through mass $A \sim 30$(^{12}C, ^{16}O, ^{20}Ne, ^{23}Na, ^{24}Mg, ^{25}Mg, ^{26}Mg, ^{27}Al, ^{29}Si, ^{30}Si and perhaps ^{31}P) in good agreement with solar system data. Explosive oxygen burning reproduces the major solar abundance features for nuclei from silicon to calcium (^{28}Si, ^{32}S, ^{33}S, ^{34}S, ^{35}Cl, ^{37}Cl, ^{36}Ar, ^{38}Ar, ^{39}K, ^{41}K, ^{40}Ca and ^{42}Ca). Finally, explosive silicon burning

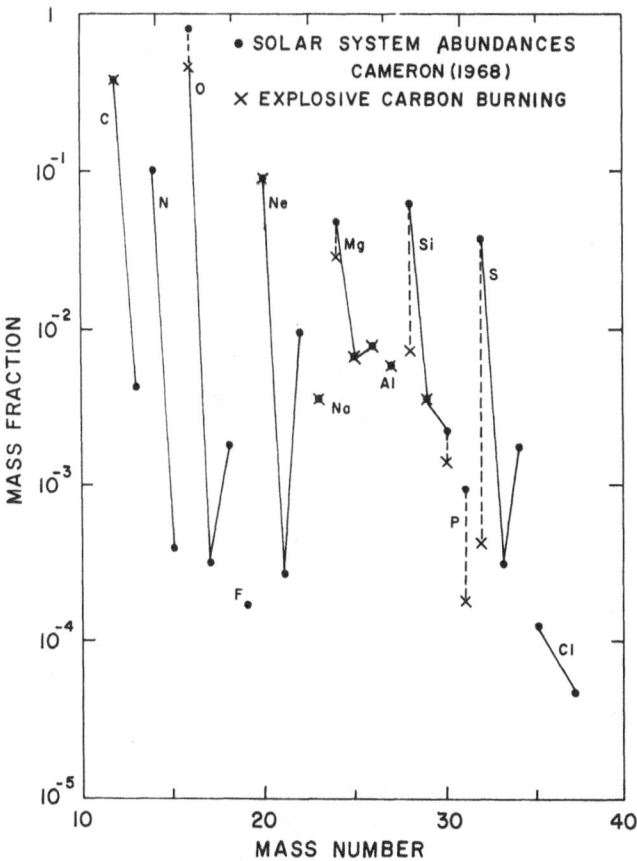

Fig. 5. The nuclear abundances by mass resulting from explosive carbon burning are compared to the solar system abundances; the two distributions are normalized at ^{20}Ne. The initial temperature and density were $T = 2 \times 10^9$ K and $\varrho = 10^5$ g cm^{-3}.

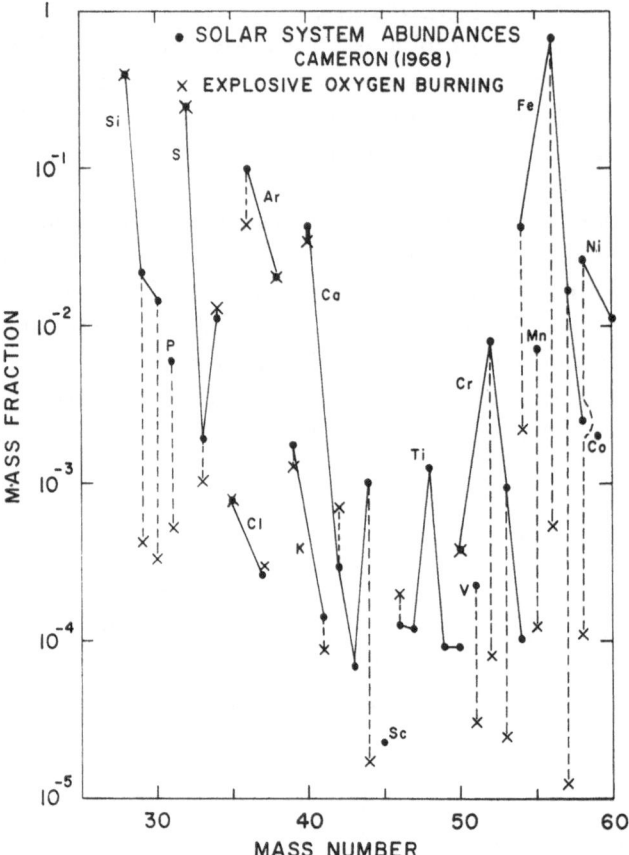

Fig. 6. The nuclear abundances by mass resulting from explosive oxygen burning are compared to the solar system abundances; the two distributions are normalized at ^{28}Si. The initial temperature and density were $T = 3.6 > 10^9$ K and $\varrho = 5 \times 10^5$ g cm^{-3}.

forms the more abundant nuclear species in the iron peak region; the abundance distributions shown in Figure 7 represents a composite (equal mass average) of two zones expanding from peak temperatures $T_9 = 4.7$ ($\varrho = 10^6$ g cm^{-3}) and $T_9 = 5.9$ ($\varrho = 2 \times 10^6$ g cm^{-3}).

One particularly encouraging feature of these explosive burning studies concerns the manner in which they reproduce the very reliably determined terrestrial isotopic abundances. The stable isotopes of chlorine, for example, are formed in these explosions as ^{35}Cl and ^{37}Ar, yet following the ^{37}Ar decay the terrestrial isotopic ratio ^{35}Cl/^{37}Cl is reproduced to better than 50 percent. This same phenomenon occurs for many other cases: the potassium isotopes ^{39}K and ^{41}K (formed as ^{41}Ca), the chromium isotopes ^{50}Cr, ^{52}Cr and ^{53}Cr (the latter two being formed *in situ* as ^{52}Fe and ^{53}Fe, respectively), the iron isotopes ^{54}Fe, ^{56}Fe and ^{57}Fe (the latter two being formed as ^{56}Ni and ^{57}Ni), and the nickel isotopes ^{58}Ni, ^{60}Ni, ^{61}Ni and ^{62}Ni (the last three being formed as ^{60}Zn, ^{61}Zn and ^{62}Zn). Furthermore, the relative abundances of ^{51}V

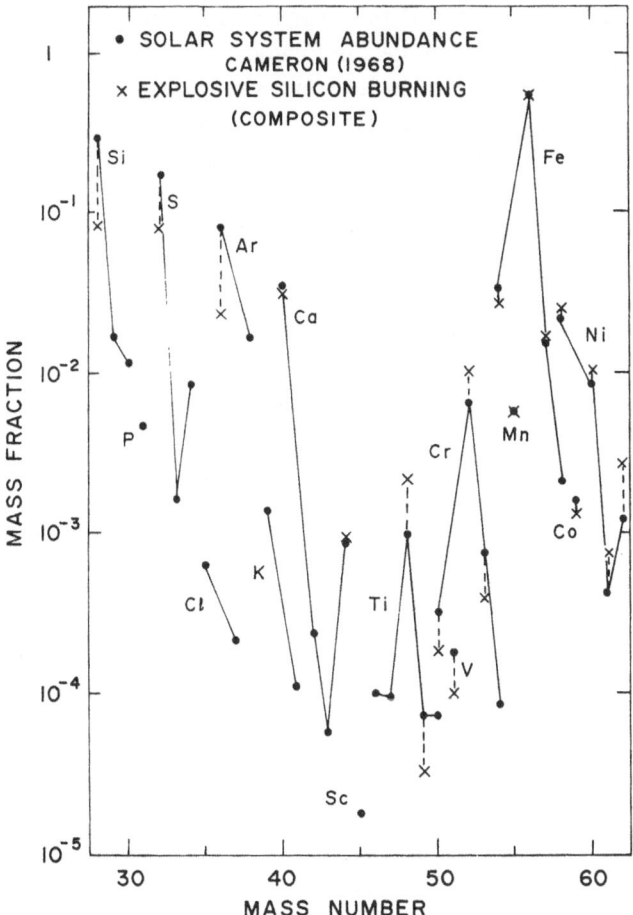

Fig. 7. The nuclear abundances by mass resulting from explosive silicon burning are compared to the solar system abundances; the two distributions are normalized at ^{56}Fe. The calculated abundances represent an equally weighted average of two mass zones burning from peak temperatures and densities $T = 4.7 \times 10^9$ K, $\varrho = 10^6$ g cm^{-3} and $T = 5.9 \times 10^9$ K, $\varrho = 2 \times 10^6$ g cm^{-3}, respectively.

(formed as ^{51}Mn), ^{55}Mn (formed as ^{55}Co) and ^{59}Co (formed as ^{59}Cu) are reproduced very well in these calculations. Assuming nature is not playing a mean trick on us, this general behavior strongly suggests that the formation of these nuclei has proceeded under explosive burning conditions.

It is important to recognize that the names of these various burning processes – explosive 'carbon', 'oxygen' and 'silicon' burning – are carryovers from the corresponding nuclear burning stages in stellar interiors. The feature which distinguishes these three *explosive* processes is the peak shocked temperature rather than the available fuel. It is quite likely that these processes take place simultaneously, but in different mass zones, in a supernova event. The thermonuclear explosion of a post helium-burning core, composed primarily of ^{12}C and ^{16}O, might naturally give rise to the required range of temperatures. If this happens, the ejected material will consist of a mix of the three distributions shown in Figures 5–7, and may reproduce in a single event the major abundance features in the mass range $20 \lesssim A \lesssim 60$.

Detailed studies of the advanced stages of evolution of more massive stars are required in order to determine whether the temperature-density structure achieved following the explosion of these fuels is consistent with the nucleosynthesis conditions explored above. In this regard, calculations of charged particle nucleosynthesis in the context of two existing supernova models are now available. Arnett et al. (1971) and Bruenn (1972a) have demonstrated that the core material ejected in ^{12}C detonation supernovae (triggered by carbon ignition taking place in an electron-degenerate core) will be composed almost exclusively of iron-peak nuclei, though considerable doubt now exists that this detonation mechanism can even result in mass ejection.* Perhaps more relevant, recent unpublished calculations by Dr Z. Barkat and the author indicate that the range of temperature-density conditions resulting from the explosion of massive carbon-oxygen cores (wherein dynamic instability is triggered by the creation of electron-positron pairs (Fraley, 1968; Barkat et al., 1967)) is generally consistent with the formation of nuclei from neon to iron by charged particle thermonuclear reactions.

5.2. r-PROCESS

Interest in the r-process mechanism has recently been stimulated by the need for more accurate predictions of the important cosmochronological production ratios ^{232}Th/^{238}U, ^{235}U/^{238}U, ^{244}Pu/^{238}U and ^{129}I/^{127}I (D. N. Schramm, these proceedings) and by questions concerning its possible role in the production of superheavy nuclei. There are considerable uncertainties associated both with the character of the astrophysical r-process environment and with nuclear mass formula predictions of the properties of neutron-rich nuclei. We will be concerned in the following discussion only with problems concerning promising astrophysical environments of r-process synthesis. It is useful to keep in mind, however, the fact that uncertainties in the nuclear physics alone allow both latitude in the choice of cosmochronological production ratios (Seeger and Schramm, 1970) and completely conflicting views as to whether superheavy nuclei can be formed by neutron-capture synthesis (Schramm and Fowler, 1971; Nix, 1972).

In an early study of the r-process mechanism (Seeger et al., 1965) a constant temperature and neutron flux were assumed, defining a steady flow path on the neutron-rich side of the valley of beta stability. Competition between neutron-capture and photoneutron reactions, for the specified conditions, determines the distribution of nuclei along each isotope chain while the beta-decay lifetimes of the more neutron-rich nuclei determine the rate of buildup to higher Z. For the large neutron fluxes characteristic of the r-process, the closed neutron shells at $N = 82$ and 126 are encountered in

---

* Additional references pertaining to the 'carbon detonation' supernova mechanism: Arnett (1969b); Bruenn (1971, 1972b); Barkat (1971); Barkat et al. (1972); Paczynski (1970, 1972); Sackmann and Weidemann (1972).

the neutron-rich regions at lower proton numbers, and the subsequent decays follow-ing neutron exhaustion form the abundance features at $A \sim 130$ and 195 (approximately 10 masses lower than the corresponding shell features at $A \sim 138$ and 208 attributed to the s-process). Seeger *et al.* concluded that two distinct astrophysical environments (or choices of conditions) are demanded to form these two features.

Cameron *et al.* (1970) have argued that a proper dynamical study of r-process synthesis must be based on the conditions expected in those types of supernovae for which hydrodynamic calculations predict a neutron-rich environment in which this process can occur. One supernova mechanism – the neutrino energy-transport mechanism – originally proposed and studied by Colgate and White (1966) predicts very promising conditions for r-process synthesis. In this model the collapse of a stellar core to near nuclear densities is accompanied by the transport of released gravitational potential energy away from the core by neutrinos and antineutrinos. Due to the extremely high densities existing in the surrounding regions, these neutrinos were able to interact with the infalling matter, heating it sufficiently to expand the outer part of the infalling core in a strong supernova shock wave which expelled the

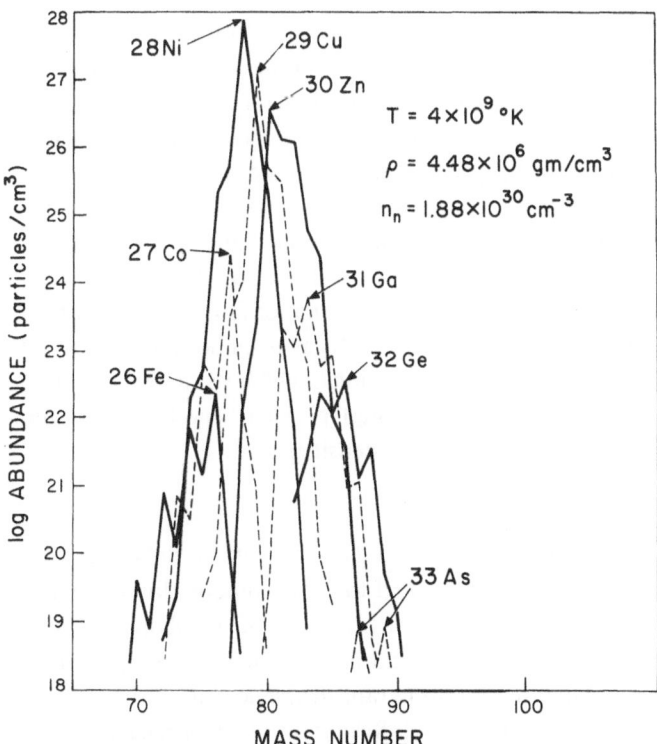

Fig. 8. The abundances of nuclear species in nuclear statistical equilibrium for neutron-rich matter ($N/Z = 8$) at high temperatures (Truran *et al.*, 1968). The most abundant heavy nucleus is ^{78}Ni. Such configurations are predicted to result in the expansion of neutronized matter from high temperatures in supernova events.

exterior mass regions into space. Several detailed calculations (Truran *et al.*, 1968; Delano and Cameron, 1971; Arnett and Truran, 1970) have established that r-process nucleosynthesis conditions can be achieved in some fraction of the ejected core material. The equilibrium abundances of nuclei at a temperature $T=4\times10^9$ K predicted for one representative expansion adiabat defined by $\varrho=7\times10^4\,T_9^3$ are illustrated in Figure 8. The three most abundant nuclei, $-\,^{78}$Ni, ^{79}Cu and ^{80}Zn $-$ all lie deep in the neutron-rich regions and are characterized by a magic number of neutrons ($N=50$). The total ratio of neutrons to protons was $N/Z=8$, and the predicted free neutron number density was enormous, $n_n=1.88\times10^{30}$ cm^{-3}. The subsequent nuclear evolution of this expanding matter has been demonstrated in the dynamical r-process calculations of Cameron *et al.* to form heavy nuclei through the transuranic region, although detailed predictions of relative *r*-process abundances were not attempted in these preliminary studies.

Recent theoretical studies raise grave doubt as to whether any significant mass ejection can occur by this neutrino transport process. Calculations by Arnett (1967) indicated that no explosion would take place for cores of mass $M>4M_\odot$, while those of Wilson (1971) predict mass ejection cannot occur for core masses $M>1.25\,M_\odot$. While these results are discouraging, there is another interesting mechanism described recently by Leblanc and Wilson (1970) which predicts comparable nucleosynthesis conditions for expanding neutronized matter. These authors have carried out a two-dimensional hydrodynamic calculation of the collapse of a star, allowing for the presence both of rotation and of magnetic fields in the interior. They found that an initially co-rotating star developed differential rotation upon collapse; consequently the rotational shear in the inner part of the resulting disk wrapped the magnetic field lines into a very tight spiral, creating an enormous magnetic energy close to the central axis of the collapsed star. The excess magnetic pressure then caused an expansion of the matter along the axis which, owing to the resulting buoyancy with respect to the local neighborhood, caused the ejection of jets of material along the axis of rotation. Leblanc and Wilson estimated that these jets contained roughly $10^{-2}\,M_\odot$ of matter which very likely has been neutronized in previous compression to very high densities.

5.3. p-PROCESS

The **p**-process or 'bypassed' nuclei, lying on the neutron-deficient side of the valley of beta stability, are generally attributed to some combination of (p,γ), (p,n) and $\gamma,n)$ reactions taking place in a proton-rich region and involving the products of earlier neutron-capture synthesis. Investigations of the p-process mechanism have been conducted by a number of authors (Frank-Kamenetskii, 1961; Ito, 1961; Arnould and Brihaye, 1969). A likely site for the formation of these nuclei is in the passage of a supernova shock wave through the hydrogen-rich outer envolope of the presupernova star. Limiting conditions consistent with p-process synthesis have recently been inferred from calculations of the appropriate rates of proton capture and neutron photodisintegration reactions (Truran and Cameron, 1972). Temperatures in excess of

2×10^9 K are required, for proton mass densities $\sim 10^2$ g cm^{-3}, if significant production of p-process nuclei is to take place on a hydrodynamic time scale. Temperatures $T \gtrsim 2 \times 10^9$ K are sufficient to assure both the rapid addition of protons onto any lighter ($A \lesssim 140$) s-process or r-process heavy elements present in these regions and the photodisintegration of any heavier neutron-capture products toward the neutron-deficient region.

6. Formation of Lithium, Beryllium and Boron

Early theories of the formation of lithium, beryllium and boron were primarily concerned with their production, together with ^{2}D, ^{3}He and ^{4}He, by spallation reactions on the surfaces of T Tauri stars (Fowler et al., 1955; Bernas et al., 1967). Such theories were criticized on very general grounds by Ryter et al. (1970) who noted that the enormous energy requirements were inconsistent with a stellar origin. The important role played by galactic cosmic rays in the formation of these nuclei was established by the calculations of Reeves et al. (1970) and Mitler (1970). These studies indicated that substantial abundances of lithium, beryllium and boron, but not of deuterium and ^{3}He, can result from the interaction of cosmic ray protons and α-particles with the more abundant heavy constituents of the interstellar medium (^{12}C, ^{14}N, ^{16}O, ^{20}Ne and ^{4}He) over the lifetime of the Galaxy.

A more thorough examination of this model was recently presented by Meneguzzi et al. (1971). Making use of the many new and improved measurements of the relevant spallation cross sections (the reader is referred to their paper for a host of references), they have calculated the rates of formation of the light elements in the context of a diffusion model for the propagation of galactic cosmic rays. Assuming a constant rate of formation of these light elements throughout galactic history, they conclude that the production of amounts of ^{6}Li, ^{9}Be, ^{10}B and ^{11}B in quantitative agreement with solar system abundances follows naturally from this mechanism. The contributions from these galactic cosmic ray interactions represent only 10% of the observed ^{7}Li abundance, however, and an inconsequential fraction ($\sim 10^{-3}$) of the observed abundances of ^{2}D and ^{3}He.

It seems reasonable to suppose that variations in the parameters of these models attributable to galactic evolutionary effects can influence the detailed predictions of the light element abundances. We might expect, for example: (1) variations in the cosmic ray flux intensity over galactic history, (2) changes in the composition of the interstellar medium, (3) destruction of these light elements in matter processed through stellar interiors, and perhaps even (4) variations in the source spectra of cosmic rays. As the calculations of Meneguzzi et al. do not consider such effects, it is of interest to ask whether their inclusion would significantly alter this picture.

Truran and Cameron (1971) have constructed and evolved models of the Galaxy in which the integrated influence of stellar and supernova nucleosynthesis on the composition of the interstellar gas is traced numerically. These models provided information concerning variations in the supernova rate and the composition of the interstellar gas over the history of the Galaxy. Assuming supernovae to be the major

source of cosmic rays, they related the production rates P_i at some time t in the past
to the current production rates in the following manner

$$P_i(t) = P_i(\text{NOW}) \left(\frac{\text{Supernova Rate at time } t}{\text{Supernova Rate NOW}} \right) \left(\frac{\chi_j(t)}{\chi_j(\text{NOW})} \right)$$

where the χ_j are abundances of the appropriate target nuclei in the interstellar medium.
It is important to distinguish between the C,N,O and Ne target abundances, which
rise from zero to their solar system values over the Galaxy's history, and that of ^{4}He,
which varies little from its assumed primordial value of 19 percent by mass; this can
influence the production of beryllium and boron (formed only by interactions involving C,N,O and Ne targets) relative to the lithium isotopes (formed by $\alpha + \alpha$ reactions
as well). Calculations performed with the production rates adjusted in this manner,
assuming constant source composition, and including the effects of the destruction
of these light elements in matter processed through stars, gave results compatible
with those of Meneguzzi et al.: the resulting abundances of ^{6}Li, ^{9}Be, ^{10}B and ^{11}B
were roughly consistent with their solar system values, but the ^{7}Li abundance was
too low.

One appealing feature of the picture which emerges from these studies is that
galactic cosmic ray interactions are capable of producing just those light nuclei
(^{6}Li, ^{9}Be, ^{10}B and ^{11}B) which cannot be formed cosmologically. ^{7}Li presents some
problems, but the calculations of Truran and Cameron (1971) indicated that, for
reasonable assumptions, it can be formed in sufficient quantities in red giant envelopes
(Cameron and Fowler 1971). These conclusions are consistent with those recently
arrived at in a more comprehensive review of problems concerning the origin of the
light elements by Reeves et al. (1973).

7. Nova Nucleosynthesis

The possible role of nova explosions in nucleosynthesis has until recently been
largely ignored. Detailed hydrodynamic studies of one promising nova mechanism
are now available which indicate that significant abundances of ^{13}C, ^{15}N, ^{17}O and
perhaps ^{19}F may be present in nova ejecta. In this model, thermonuclear runaways
occurring in hydrogen-rich envelopes accreted on the surfaces of hot white dwarfs are
found to give rise to direct shock wave ejection of some fraction of the envelope
(Rose and Smith, 1971; Starrfield et al., 1972). The mass accretion mechanism is
strongly suggested by observational studies of old novae, which show them to be
close binary systems consisting of a small hot white dwarf and a larger, cooler star
which appears to be shedding hydrogen-rich material (Kraft, 1964; Paczynski, 1965;
Mumford, 1967).

Hydrogen burning under explosive high-temperature conditions proceeds via
CNO-cycle reactions. Starrfield et al. have followed these reactions in detail through
the entire hydrodynamic history of one of these events. The resulting abundances

differ dramatically from those predicted for CNO-burning in stellar cores on time scales $\sim 10^6$–10^9 yr (see Section 4). The *hydrodynamic* times scales are typically *less* than the positron-decay lifetimes of the unstable nuclei which participate in these cycles (^{13}N(870 s), ^{15}O(178 s) and ^{17}F(95 s), see Table III), hence these nuclei tend to accumulate. Their subsequent decay gives rise to large concentrations of the stable isotopes ^{13}C, ^{15}N and ^{17}O relative to ^{12}C, ^{14}N and ^{16}O in the ejected material. For

TABLE III

Carbon-nitrogen-oxygen
(CNO) Bi-cycle

$$^{12}C + {}^{1}H \rightarrow {}^{13}N + \gamma$$
$$^{13}N \rightarrow {}^{13}C + e^{+} + \nu$$
$$^{13}C + {}^{1}H \rightarrow {}^{14}N + \gamma$$
$$^{14}N + {}^{1}H \rightarrow {}^{15}O + \gamma$$
$$^{15}O \rightarrow {}^{15}N + e^{+} + \nu$$
$$^{15}N + {}^{1}H \rightarrow {}^{12}C + {}^{4}He + \gamma$$

or

$$^{15}N + {}^{1}H \rightarrow {}^{16}O + \gamma$$
$$^{16}O + {}^{1}H \rightarrow {}^{17}F + \gamma$$
$$^{17}F \rightarrow {}^{17}O + e^{+} + \nu$$
$$^{17}O + {}^{1}H \rightarrow {}^{14}N + {}^{4}He + \gamma$$

the range of models studied by Starrfield *et al.* the isotope mass ratios averaged over the nova ejecta were $^{13}C/^{12}C \sim 2$–7, $^{15}N/^{14}N \sim 1.5$–2.5 and $^{17}O/^{16}O \sim 0.04$–0.4 as compared to the solar system ratios 0.013, 0.0044 and 0.00042, respectively.

The ratios achieved represent consistent isotopic enhancements of factors ~ 200 to 400 for these three cases. Recent studies by Dr S. Starrfield and the author suggest that significant ^{19}F production can also occur, but it is not yet clear whether a comparable overabundance can be achieved. The potential importance of these results lies in the fact that these four nuclei are not readily formed in other astrophysical environments. Howard, Arnett and Clayton (1971) have suggested that the formation of some of these isotopes together with ^{18}O, ^{21}Ne and ^{22}Ne might take place in the passage of shock waves through helium zones in supernova envelopes, but they were unable to account for ^{13}C and ^{17}O by this mechanism. The nova mechanism seems most promising, but further studies are required before any firm conclusions can be drawn.

8. Summary and Discussion

Future research on the various mechanisms of nucleosynthesis described in this review is certain to provide surprises which will demand modifications of existing theories. The overall picture which emerges from a summary of current nucleosynthesis theories is nonetheless quite encouraging:

8.1. THE LIGHT ELEMENTS

Hydrogen, ^{2}D, ^{3}He and ^{4}He represent the major products emerging from the cosmological big-bang; one interesting universe (defined by $T_0 = 2.7$ K, $\varrho_0 \sim 3 \times 10^{-31}$ g cm^{-3} and zero leptonic number) predicts relative abundances of these nuclei consistent with solar system abundances. The observed abundances of ^{6}Li, ^{9}Be, ^{10}B and ^{11}B are consistent with the contributions predicted to result from the interaction of cosmic rays with the interstellar medium, integrated over the history of the galaxy. The ^{7}Li abundances predicted by both these mechanisms are low by factors ~ 5 to 10; Cameron and Fowler (1971) have demonstrated that significant ^{7}Li production might be expected to occur as a consequence of mixing in red giant envelopes. The origin of all nuclei of mass $A < 12$ is thus accounted for.

8.2. THE ELEMENTS CARBON-TO-FLUORINE

^{12}C and ^{16}O, in roughly comparable amounts, represent the primary products of helium burning (assuming subsequent nuclear experiments confirm an intermediate value for the reduced α-particle width of the 7.12 MeV excited state in ^{16}O). ^{14}N is formed in CNO-cycle hydrogen burning; this very likely takes place in the hydrogen burning shells of red giant stars where the ^{14}N thus formed can be transported to the surface by convection and enrich the interstellar medium as a consequence of the mass loss observed to occur in these stars. Nova explosions formed enormously enhanced abundances of ^{13}C, ^{15}N and ^{17}O in relative amounts consistent with solar system abundances; further calculations are required to determine whether comparable overabundances of ^{19}F and perhaps ^{21}Ne can be achieved in such events. ^{18}O, ^{21}Ne and ^{22}Ne may be formed in the flashing of helium zones (^{14}N$(\alpha, \gamma)^{18}$F$(e^+ \nu)^{18}$O followed by ^{18}O$(\alpha, n)^{21}$Ne and ^{18}O$(\alpha, \gamma)^{22}$Ne, but detailed studies are required to confirm this mechanism.

8.3. THE ELEMENTS NEON-TO-NICKEL

The explosive burning of ^{12}C and ^{16}O fuels at temperatures in the range $2 \lesssim T_9 \lesssim 7$, identified with some type of supernova event, will form most of the nuclei from ^{20}Ne to ^{62}Ni. The mode of production of the less abundant neutron-rich isotopes – ^{36}S, ^{46}Ca, ^{48}Ca, ^{50}Ti, ^{54}Cr, ^{58}Fe and ^{64}Ni – requires special attention (Howard et al., 1972; Peters et al., 1972; Truran, 1972). Unfortunately, calculations of explosive nucleosynthesis do not provide a reliable prediction of the cosmochronologically important source production ratio ^{40}K$/^{40}$Ar. Existing nuclear chronologies record only the history of r-process synthesis in the galaxy and it would be particularly interesting to know whether these explosive burning processes have a parallel history.

8.4. THE HEAVY ELEMENTS

The mechanisms originally proposed by Burbidge et al. (1957) – the s-, r-, and p-processes – explain very well the major abundance features in the heavy element region ($A > 60$). The precise conditions under which these processes occur remain uncertain, but it does seem clear that r-process and p-process synthesis proceed in explosive environments (supernovae). The most promising site for s-process synthesis is red giant envelopes, where helium-shell-flashing phenomenon can provide an appropriate neutron flux.

One seems compelled to draw one very general conclusion from the studies reviewed in this paper: *nature prefers explosive environments as the sites of element synthesis.* Nuclear processes are clearly essential to meet the energy requirements of extended stable phases of stellar evolution, but severe difficulties are encountered when one tries to explain how the product nuclei can be removed unscathed from stellar cores. In contrast, the primeval fireball, supernova explosions and nova explosions provide not only appropriate nucleosynthesis conditions but also very natural explanations of the presence of the products of their nuclear burning in the interstellar gas. Even the s-process, driven by relatively *weak* neutron fluxes, seems to require flashing phenomenon in helium shells. Finally, the cosmic rays protons and α-particles which interact with the interstellar gas to form lithium, beryllium and boron very likely are accelerated to cosmic ray energies by shock phenomenon in the envelopes of supernovae.

This research was supported in part by the National Science Foundation under grant GP-30289. The author wishes to thank Dr R. Jastrow for the hospitality of the Goddard Institute for Space Studies where various calculations relevant to his own research have been performed.

References

Abraham, Z. and Iben, I., Jr.: 1971, *Astrophys. J.* **170**, 157.
Allen, B. J., Gibbons, J. H., and Macklin, R. L.: 1972, preprint.
Alpher, R. A., Bethe, H. A., and Gamov, G.: 1948, *Phys. Rev.* **73**, 803.
Alpher, R. A. and Herman, R. C.: 1950, *Rev. Mod. Phys.* **22**, 153.
Arnett, W. D.: 1967, *Can. J. Phys.* **45**, 1621.
Arnett, W. D.: 1969a, *Astrophys. J.* **157**, 1369.
Arnett, W. D.: 1969b, *Astrophys. Space Sci.* **5**, 180.
Arnett, W. D. and Truran, J. W.: 1969, *Astrophys. J.* **157**, 339.
Arnett, W. D. and Truran, J. W.: 1970, *Astrophys. J.* **160**, 959.
Arnett, W. D., Truran, J. W., and Woosley, S. E.: 1971, *Astrophys. J.* **165**, 87.
Arnould, M. and Brihaye, C.: 1969, *Astron. Astrophys.* **1**, 193.
Austin, S. M., Trentelman, G. F., and Kasky, E.: 1971, *Astrophys. J.* **163**, L79.
Bahcall, J. N. and Ulrich, R. K.: 1971, *Astrophys. J.* **170**, 593.
Barkat, Z.: 1971, *Astrophys. J.* **163**, 433.
Barkat, Z., Rakavy, G., and Sack, N.: 1967, *Phys. Rev. Letters* **18**, 379.
Barkat, Z., Wheeler, J. C., and Buchler, J.-R.: 1972, *Astrophys. J.* **171**, 651.
Bernas, R., Gradsztajn, E., Reeves, H., and Schatzman, E.: 1967, *Ann. Phys.* **44**, 426.
Bethe, H. A.: 1939, *Phys. Rev.* **55**, 434.
Bodansky, D., Clayton, D. D., and Fowler, W. A.: 1968, *Astrophys. J. Suppl. No. 148* **16**, 299.
Bruenn, S. W.: 1971, *Astrophys. J.* **168**, 203.
Bruenn, S. W.: 1972a, *Astrophys. J. Suppl. No. 207* **24**, 283.
Bruenn, S. W.: 1972b, 'The Effect of Urca Shells on the Density of Carbon Ignition in Degenerate Stellar Cores', preprint.
Burbidge, E. M., Burbidge, G. R., Fowler, W. A., and Hoyle, F.: 1957, *Rev. Mod. Phys.* **29**, 547.
Cameron, A. G. W.: 1957, Chalk River Report CRL-41.
Cameron, A. G. W.: 1963, *Nuclear Astrophysics*, Yale University lecture notes.

Cameron, A. G. W.: 1968, in L. H. Ahrens (ed.), *The Origin and Distribution of the Elements*, Pergamon Press, New York.

Cameron, A. G. W., Delano, M. D., and Truran, J. W.: 1970, in *Proceedings of the International Conference on the Properties of Nuclei far from the Valley of Beta Stability*, Vol. 2, CERN.

Cameron, A. G. W. and Fowler, W. A.: 1971, *Astrophys. J.* **164**, 111.

Caughlan, G. R.: 1965, *Astrophys. J.* **141**, 688.

Clayton, D. D., Fowler, W. A., Hull, T. E., and Zimmerman, B. A.: 1961, *Ann. Phys.* **12**, 331.

Climenhaga, J. L.: 1957, *Pub. Dom. Ap. Obs.* **11**, 307.

Colgate, S. A. and White, R. H.: 1966, *Astrophys. J.* **143**, 626.

Danziger, I. J.: 1965, *Monthly Notices Roy. Astron. Soc.* **131**, 51.

Davis, R., Jr.: 1955, *Phys. Rev.* **97**, 766.

Davis, R., Jr.: 1972, private communication.

Davis, R., Jr., Rogers, L. C., and Radeka, V.: 1971, *Bull. Am. Phys. Soc.* **16**, 631.

Deinzer, W. and Salpeter, E. E.: 1964, *Astrophys. J.* **140**, 499.

Delano, M. D. and Cameron, A. G. W.: 1971, *Astrophys. Space Sci.* **10**, 203.

Dworetsky, M. M., Ross, J. E., and Aller, L. H.: 1970, *Bull. A.A.S.* **2**, 311.

Fowler, W. A.: 1972, 'What Cooks with Solar Neutrinos', preprint.

Fowler, W. A. and Hoyle, F.: 1964, *Astrophys. J. Suppl. No. 91* **9**, 201.

Fowler, W. A., Burbidge, E. M., and Burbidge, G. R.: 1955, *Astrophys. J.* **122**, 271.

Fraley, G. S.: 1968, *Astrophys. Space Sci.* **2**, 96.

Frank-Kamenetskii, D. A.: 1961, *Soviet Astron. –AJ* **5**, 66.

Gari, M. and Huffman, A. H.: 1972, 'Interaction Contributions to the Solar p-p Reaction', preprint.

Geiss, J. and Reeves, H.: 1972, *Astron. Astrophys.*, in press.

Gibbons, J. H. and Macklin, R. L.: 1968, in W. D. Arnett, C. J. Hansen, J. W. Truran and A. G. W. Cameron (eds.), *Nucleosynthesis*, Gordon and Breach, New York.

Greenstein, G. S.: 1968, *Astrophys. Space Sci.* **2**, 155.

Hansen, C. J.: 1971, *Astrophys. Space Sci.* **14**, 389.

Hansen, C. J. and Zaidins, C. S.: 1971, *Astrophys. J.* **168**, 317.

Howard, W. M., Arnett, W. D., and Clayton, D. D.: 1971, *Astrophys. J.* **165**, 495.

Howard, W. M., Arnett, W. D., Clayton, D. D., and Woosley, S. E.: 1972, *Astrophys. J.* **175**, 201.

Hoyle, F.: 1946, *Monthly Notices Roy. Astron. Soc.* **106**, 343.

Ito, K.: 1961, *Progr. Theoret. Phys.* **26**, 990.

Jaszczak, R. J. and Macklin, R. L.: 1970, *Phys. Rev.* **C2**, 2452.

Jaszczak, R. J., Gibbons, J. H., and Macklin, R. L.: 1970, *Phys. Rev.* **C2**, 63.

Kraft, R. P.: 1964, *Astrophys. J.* **139**, 457.

Leblanc, J. M. and Wilson, J. R.: 1970, *Astrophys. J.* **161**, 541.

Macklin, R. L. and Gibbons, J. H.: 1965, *Rev. Mod. Phys.* **37**, 166.

Mazarakis, M. G. and Stephens, W. E.: 1972, *Astrophys. J.* **171**, L97.

McCaslin, S. J., Mann, F. M., and Kavanagh, R. W.: 1972, 'The Mass of the Second Excited State of ^{12}C', preprint.

Meneguzzi, M., Audouze, J., and Reeves, H.: 1971, *Astron. Astrophys.* **15**, 337.

Merrill, P. W.: 1952, *Science* **115**, 484.

Michaud, G. J.: 1972, *Astrophys. J.* **175**, 751.

Michaud, G. and Fowler, W. A.: 1972, *Astrophys. J.* **173**, 157.

Michaud, G. J. and Vogt, E. W.: 1972, *Phys. Rev.* **C5**, 350.

Mitler, H. E.: 1970, 'Smithsonian Astrophysical Observatory', Special Report No. 330.

Mumford, G. S.: 1967, *Publ. Astron. Soc. Pacific* **79**, 283.

Nix, J. R.: 1972, *Ann. Rev. Nucl. Sci.* **22**, in press.

Paczynski, B.: 1965, *Acta Astron.* **15**, 197.

Paczynski, B.: 1970, *Acta Astron.* **20**, 47.

Paczynski, B.: 1972, *Astrophys. Letters* **11**, 53.

Patterson, J. R., Winkler, N., and Zaidins, C. S.: 1969, *Astrophys. J.*, **157**, 367.

Peebles, P. J. E.: 1966a, *Phys. Rev. Letters* **16**, 410.

Peebles, P. J. E.: 1966b, *Astrophys. J.* **146**, 542.

Penzias, A. A. and Wilson, R. W.: 1965, *Astrophys. J.* **142**, 419.

Peters, J. G.: 1968, *Astrophys. J.* **154**, 224.

Peters, J. G., Fowler, W. A., and Clayton, D. D.: 1972, *Astrophys. J.* **173**, 637.

Preston, G. W.: 1971, *Astrophys. J.* **164**, L41.

Puhlhofer, F., Ritter, H. G., Bock, R., Brommundt, G., Schmidt, H., and Bethge, K.: 1970, *Nucl. Phys.* **A147**, 258.

Rakavy, G., Shaviv, G., and Zinamon, Z.: 1967, *Astrophys. J.* **150**, 131.

Reeves, H., Fowler, W. A., and Hoyle, F.: 1970, *Nature* **226**, 727.

Reeves, H., Audouze, J., Fowler, W. A., and Schramm, D. N.: 1973, *Astrophys. J.* **179**, 909.

Rose, W. K. and Smith, R. L.: 1972, *Astrophys. J.* **172**, 699.

Ryter, C., Reeves, H., Gradsztajn, E., and Audouze, J.: 1970, *Astron. Astrophys.* **8**, 389.

Sackmann, J. and Weidemann, V.: 1972, *Astrophys. J.* **178**, 427.

Salpeter, E. E.: 1952, *Astrophys. J.* **115**, 326.

Sanders, R. H.: 1967, *Astrophys. J.* **150**, 971.

Schmidt, M.: 1963, *Astrophys. J.* **137**, 758.

Schramm, D. N. and Fowler, W. A.: 1971, *Nature* **231**, 103.

Schwarzschild, M. and Harm, R.: 1967, *Astrophys. J.* **150**, 961.

Seeger, P. A., Fowler, W. A., and Clayton, D. D.: 1965, *Astrophys. J. Suppl. No. 97* **11**, 121.

Seeger, P. A. and Schramm, D. N.: 1970, *Astrophys. J.* **160**, L157.

Spinka, H. and Winkler, H.: 1972, *Astrophys. J.* **174**, 455.

Starrfield, S., Truran, J. W., Sparks, W. M., and Kutter, G. S. 1972, *Astrophys. J.* **176**, 169.

Stephens, W. E. and Mazarakis, M. G.: 1970, *Bull. A. P. S.* **15**, 629.

Truran, J. W.: 1972, *Astrophys. J.* **177**, 453.

Truran, J. W. and Arnett, W. D.: 1970, *Astrophys. J.* **160**, 181.

Truran, J. W. and Arnett, W. D.: 1971, *Astrophys. Space Sci.* **11**, 430.

Truran, J. W. and Cameron, A. G. W.: 1971, *Astrophys. Space Sci.* **14**, 179.

Truran, J. W. and Cameron, A. G. W.: 1972, *Astrophys. J.* **171**, 89.

Truran, J. W., Cameron, A. G. W., and Gilbert, A.: 1966, *Can. J. Phys.* **44**, 563.

Truran, J. W., Arnett, W. D., and Cameron, A. G. W.: 1967, *Can. J. Phys.* **45**, 2315.

Truran J. W., Arnett, W. D., Tsuruta, S., and Cameron, A. G. W.: 1968, *Astrophys Space Sci.* **1**, 129.

Vidal, N. V., Shaviv, G., and Kozlovsky, B.-Z.: 1971, *Astron. Astrophys.* **13**, 147.

von Weizsacker, C. F.: 1938, *Physik. Z.* **39**, 633.

Wagoner, R. V.: 1968, *Astrophys. J.* **151**, L103.

Wagoner, R. V.: 1969, *Astrophys. J. Suppl. No. 162* **18**, 247.

Wagoner, R. V.: 1973, *Astrophys. J.* **179**, 343.

Wagoner, R. V., Fowler, W. A., and Hoyle, F.: 1967, *Astrophys. J.* **148**, 3.

Wallerstein, G. and Greenstein, J. L.: 1964, *Astrophys. J.* **139**, 1163.

Wallerstein, G., Green, T. F., and Tomley, L. J.: 1967, *Astrophys. J.* **150**, 245.

Weisser, D. C., Morgan, J. F., and Thompson, D. R.: 1971, unpublished.

Wilson, J. R.: 1971, *Astrophys. J.* **163**, 209.

Woosley, S. E., Arnett, W. D., and Clayton, D. D.: 1972, *Astrophys. J.* **175**, 731.

Wyller, A. A.: 1965, *Astrophys. J.* **143**, 829.

NUCLEO-COSMOCHRONOLOGY

DAVID N. SCHRAMM

Calif. Inst. of Technology Pasadena, Calif., U.S.A.,
Inst. of Theor. Astron. Cambridge, England

and

Univ. of Texas PMA 16.228* Austin, Tex., U.S.A.

Abstract. A review is made of the basis of nucleo-cosmochronologies, emphasizing the model-independent approach. The exponential model as well as recent galactic evolution models are discussed. The r-process chronometer pairs ^{252}Th/^{238}U, ^{235}U/^{238}U, ^{244}Pu/^{232}Th and ^{129}I/^{127}I are discussed in detail. The possible development of other chronometer-pairs is also discussed. In particular it is shown that the ^{187}Re–^{187}Os system may eventually be able to determine the mean age of the elements to higher precision than any other r-process chronometer because all of the parameters are experimentally measurable. The possibility of a p-process chronology based on ^{146}Sm and an s-process chronology from ^{176}Lu is also examined.

Nucleo-cosmochronology is the use of the relative abundances of radioactive nuclei to determine the timescales for the nucleosynthesis of these nuclei. Since the nucleosynthetic processes producing these nuclei presumably took place in stars in the galaxy, these chronologies yield time scales for the age of the galaxy, and thus a lower limit on the age of the Universe. It has been shown by Schramm and Wasserburg (1970) that certain information is obtained independent of the model of the evolution of galaxies used. In particular the mean age of elements, and the time between the last nucleosynthetic event and the solidification of objects in the solar system can be determined model independently. It is also possible to determine some model-independent features of the variation of the rate of nucleosynthesis with time. The actual chronometers used has increased in the past few years and this increase has resulted in a flurry of interest in the subject. Burbidge *et al.* (1957) and Fowler and Hoyle (1960) used the pairs ^{232}Th/^{238}U and ^{235}U/^{238}U to develop model dependent chronologies for galactic nucleosynthesis. Hohenberg *et al.* (1967) found that ^{129}Xe from the decay of ^{129}I $(\tau_{1/2} \sim 17 \times 10^6$ yr) yielded a sharp isochronism in the forma- of chrondritic meteorites. Then Wasserbrug *et al.* (1969) found that fission like Xe gas and excess fission tracks were correlated in the meteorite St. Severin indicating a fissioning nucleus as the origin; the most likely candidate being ^{244}Pu. This was subsequently proved by Alexander *et al.* (1971) who obtained the Xe fission yield spectrum of ^{244}Pu and found it agreed with the fission Xe observed in St. Severin and several other meteorites. Armed with the additional chronometer pairs ^{129}I/^{127}I and ^{244}Pu/^{238}U various workers developed model-dependent nucleochronologies (Wasserburg *et al.*, 1969; Hohenberg, 1969; Fowler, 1972; Kohman, 1972). In order to understand these model-dependent results, we will use the model-independent procedure of Schramm and Wasserburg (1970) which will be described below.

* Present address.

A. G. W. Cameron (ed.), Cosmochemistry, 51–67. All Rights Reserved
Copyright © 1973 by D. Reidel Publishing Company, Dordrecht-Holland

1. General Theory

Let the rate of change of the abundance of a nuclear species N_i with decay constant λ_i be described by the following equation.

$$dN_i/d\tau = -\lambda_i N_i + P_i p(\tau) \tag{1}$$

for $0 \leqslant \tau \leqslant T$ where T is the total duration of nucleosynthesis; $P_i p(\tau)$ is the time dependent production; and the relative production of two species is $P_i p(\tau)/P_j p(\tau) = = P_i/P_j$, which is assumed constant for all sources. The various times used are illustrated in Figure 1a.

It is possible to make Equation (1) more general by adding a term representing possible dilution of the interstellar gas due to infalling matter, or possible destruction mechanisms such as trapping in remnant stars. Such a term has been included in the derivations of Schramm and Wasserburg (1970) and is found to not effect the model independent results to be discussed here. Thus, to simplify the equations such a term will not be included here. Equation (1) can also be written in terms of mass fraction $X_i = N_i A_i M_u/M_G$ and the fractional production rate $\Pi_i = y_i A M_G/A M_u$ where A_i is the atomic mass; M_u is the mass of an atomic mass unit; M_G is the mass of the interstellar gas; y_i is the newly created mass fraction of element i in material ejected from the nucleosynthetic process considered; and A is the mass ejection rate for such nucleosynthetic events. The variables X_i and Π_i have been used by Fowler (1972) because of the ease in using those parameters in stellar formation and galactic evolution models. It can be shown that such a parameter change does not effect the

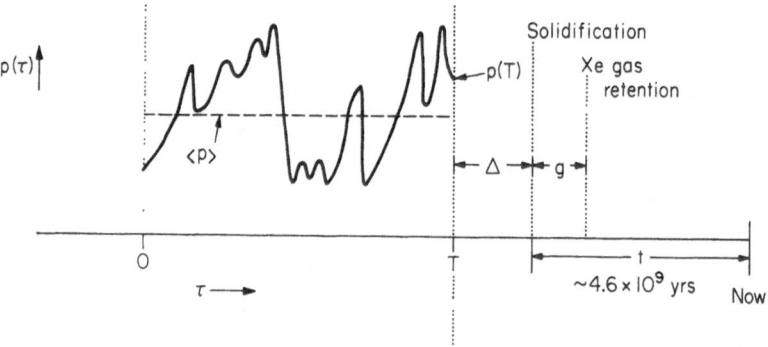

Fig. 1a. A schematic showing the meaning of the notation used. $p(\tau)$ is the rate of nucleosynthesis as a function of time starting at $\tau = 0$; T is the total duration of nucleosynthesis contributing to the solar system; $\langle p \rangle$ is the average rate of nucleosynthesis; and $p(T)$ is the rate just prior to the separation of the pre-solar system gas from the galactic gas containing new nucleosynthetic ejecta. Δ represents the separation between the last nucleosynthetic event contributing to the solar system and the time of solidification of solid bodies in the solar system; g represents the time interval between solidification and Xe gas retention (g is normally taken to be 0) and t is the age of solid bodies in the solar system ($\sim 4.6 \times 10^9$ yr.). The total age of the elements considered is $T + \Delta + t$ which for the r-process can be taken as the age of the galaxy.

nucleochronologic equations. Equation (1) can be solved for N_i at the time of solidification of solid bodies in the solar system.

$$N_i(T + \Delta) = P_i T \langle p \rangle \exp\left[-\lambda_i(T + \Delta)\right] \int_0^T \exp(\lambda_i \xi) \cdot \varrho(\xi) \, d\xi, \tag{2}$$

where $\langle p \rangle = (1/T) \int_0^T p(\xi) \, d\xi$ is the mean rate of nucleosynthesis, and $\varrho(\xi) = p(\xi)/T \langle p \rangle$ with $\int_0^T \varrho(\xi) \, d\xi = 1$ and where Δ is the time between the last nucleosynthetic event contributing to the solar system material and the time of solidification of objects in the solar system. Thus, for a stable nucleus with $\lambda_i = 0$, $N_i(T + \Delta) = P_i T \langle p \rangle$. Equation (2) can be greatly simplified in the limits $\lambda T \ll 1$ and $\lambda T \gg 1$ as will be shown.

2. Long-Lived Isotopes $\lambda T \ll 1$

If $p(\xi) \geqslant 0$ the quantity $p(\xi)$ is like a probability density function with moments μ_n about the mean.

$$\mu_n = \int_0^T (\xi - \langle \tau \rangle)^n \varrho(\xi) \, d\xi, \tag{3}$$

where $\langle \tau \rangle = \int_0^T \zeta(\xi) \, d\xi$ and $\mu_1 = 0$.

Therefore Equation (2) can be written as

$$N_i(T + \Delta) = P_i T \langle p \rangle \exp\left[-\lambda_i(\Delta + T - \langle \tau \rangle)\right] \left[1 + \zeta_i\right],$$

where

$$\zeta_i = \sum_{n=0}^{\infty} \frac{\lambda_i^{n+2} \mu_{n+2}}{(n+2)!}. \tag{4}$$

In the long-lived limit $\lambda T \ll 1$, $\zeta_i \to 0$.

Thus

$$N_i(T + \Delta) \simeq P_i T \langle p \rangle \exp\left[-\lambda_i(\Delta + T - \langle \tau \rangle)\right]. \tag{5}$$

To determine a nucleochronology, relative rates are necessary. For simplicity let

$$R(i, j) \equiv \frac{P_i/P_j}{N_i(T + \Delta)/N_j(T + \Delta)},$$

for a nucleochronologic problem this parameter $R(i, j)$ and the decay lifetimes are the input data.

Defining

$$\Delta_{ij}^{\max} = \frac{\ln[R(i, j)]}{\lambda_i - \lambda_j},$$

then for two species i and j, Equation (5) can be written as

$$T - \langle \tau \rangle = \Delta_{ij}^{\max} - \Delta \tag{6}$$

$T - \langle \tau \rangle$ is just the mean age of the elements measured backward from T. The mean age measured from the present time is just

$$T - \langle \tau \rangle + \Delta + t = \Delta_{ij}^{\max} + t,$$

where t is the age of solid bodies in the solar system, taken here to be $(4.6 \pm 0.1) \times 10^9$. Given a particular model for the distribution of nucleosynthetic sources with time, yields a model-dependent relation between $\langle \tau \rangle$ and T and thus enables the model-dependent determination of the total duration of that type of nucleosynthesis. For example if it is assumed that the rate of nucleosynthesis has been constant ($p(\tau) = \text{constant}$) then $\langle \tau \rangle = T/2$ which would yield $T \approx 2\Delta_{ij}^{\max}$ for $\lambda T \ll 1$. Note that if all nucleosynthesis occurred in a single event then $T \simeq 0$ and $\Delta = \Delta_{ij}^{\max}$ which is the maximum time separation between nucleosynthesis and solar system solidification. It is normally assumed in galactic evolution models (Truran and Cameron, 1971; Talbot and Arnett, 1972) that $T > 0$ and the rate of nucleosynthetic events occurring early was equal to or greater than the average rate $[p(0) \gtrsim \langle p \rangle]$.

This implies that $T/2 \leqslant T - \langle \tau \rangle \leqslant T$; thus for $\lambda T \ll 1$; $\Delta_{ij}^{\max} \leqslant T \leqslant 2\Delta_{ij}^{\max}$.

It is clear from the above analysis that if a nucleochronometric pair could be found in which both nuclei satisfy $\lambda T \ll 1$ and both are produced in the same nucleosynthetic process, then it would be possible to model independently determine the mean age of all elements produced in that process. At present the only long-lived chronometer pair which has been developed is $^{232}\text{Th}/^{238}\text{U}$. Although $\lambda_{232} T \ll 1$ for all reasonable models, $\lambda_{238} T$ is not. However, it has been found that the correction to Equation (6) is less than 20% for most models, thus $^{232}\text{Th}/^{238}\text{U}$ almost enables a model independent determination of the mean age of r-process material. A superior chronometer pair for this purpose would be $^{187}\text{Re}/^{187}\text{Os}$, unfortunately at the present time, there are considerable uncertainties in the half-life and $R(i, j)$ for $^{187}\text{Re}/^{187}\text{Os}$. These uncertainties will be discussed in more detail later. In connection with long lived chronometers it is important to remember that they only determine $T - \langle \tau \rangle$, and this mean age must agree for all nuclei produced in the same process, which means that $\Delta_{ij}^{\max}$ must be the same value for all nuclei with $\lambda T \ll 1$ produced in the same process.

3. Short-Lived Isotopes $\lambda T \gg 1$

It was shown by Schramm and Wasserburg that if $p(\tau)$ was smooth near $\tau = T$

$$\frac{1}{p(\tau)} \frac{dp}{d\tau}\bigg|_{\tau = T} \ll \lambda_i, \text{ then for } \lambda_i T \gg 1$$

Equation (2) reduces to

$$N_i(T + \Delta) \simeq \frac{P_i}{\lambda_i} p(T) \exp(-\lambda_i \Delta).$$ (7)

The ratio of two such short-lived isotopes yields

$$\Delta \simeq \frac{1}{\lambda_i - \lambda_j} \ln\left[R(i, j) \frac{\lambda_j}{\lambda_i}\right].$$ (8)

If $p(\tau)$ is not smooth near $\tau = T$, but has a spiked behavior for example

$$p(\tau) = p_0(\tau) + d\delta(\tau - T), \text{ then}$$

$$\Delta = \frac{1}{\lambda_i - \lambda_j} \ln\left[\frac{\lambda_j R(ij)}{\lambda_i} \cdot \frac{p_0(T) + \lambda_i d}{p_0(T) + \lambda_j d}\right].$$ (9)

Thus in general for arbitrary size spikes $(\lambda T \gg 1)$

$$\frac{1}{\lambda_i - \lambda_j} \ln\left[R(i, j) \frac{\lambda_j}{\lambda_i}\right] \leq \Delta \leq \Delta_{ij}^{\max}.$$

Thus a range is placed on Δ in a model independent manner. If a short-lived isotope is taken relative to a long-lived or stable nucleus (L), then the rate of nucleosynthesis at T, $p(T)$, can be compared with the average rate $\langle p \rangle$, since the relevant equations reduce to

$$\frac{1}{T^*} \equiv \frac{p(T)}{T\langle p \rangle} = \frac{\lambda_i}{R(i, L)} \exp[(\lambda_i - \lambda_L)\Delta][1 - (T - \langle \tau \rangle)\lambda_L].$$

Since $T - \langle \tau \rangle$ can be determined from the ratio of two long-lived nuclei, then T^* can be calculated. If $T^* < T$, then there was a higher than average rate of nucleosynthesis at $\tau = T(p(T) > \langle p \rangle)$.

The short-lived isotopes used to date are ^{129}I and ^{244}Pu, and they are usually taken relative to the stable and long-lived isotopes ^{127}I and ^{232}Th (some workers use ^{238}U; however, ^{232}Th is superior since then the dependence on T can be decoupled). These short-lived nuclei enable the model-independent determination of T^* and Δ for r-process nucleosynthesis. Another short-lived nucleus, ^{146}Sm, may yield a p-process chronometer (Audouze and Schramm, 1972).

4. Intermediate-Lived Nuclei

For nuclei with a longer life than used in the above section the expansion of Equation (7) must include higher order terms.

$$N_i(T + \Delta) \simeq \frac{P_i}{\lambda_i} p(T) \exp(-\lambda_i \Delta)\left[1 - \frac{p'(T)}{\lambda_i p(T)} \cdots\right],$$

where

$$p'(T) = \frac{dp(\tau)}{d\tau}\Bigg|_{\tau=T}.$$

Normalizing to a stable or long-lived nucleus L, yields

$$1 = R(i, L)\frac{\exp\left[-(\lambda_i - \lambda_L)\varDelta\right]}{\lambda_i T^*}\left[1 - \frac{p'(T)}{\lambda_i p(T)} \cdots\right].$$

Thus if T^* is known from the short-lived nucleus, then intermediate nuclei enable the determination of the derivatives of $p(\tau)$ about T. If several intermediate nuclei were available then higher order derivatives could be obtained, and the shape of $p(\tau)$ could be determined. At present the only intermediate nucleus is ^{235}U. The use of this nucleus enables the determination of $p'(T)/p(T)$.

When a particular model is assumed then $p'(T)$ is defined by the model. In order for that model to be valid, requires the model $p'(T)$ to be consistent with the $p'(T)$ determined from ^{235}U model independently.

Short- and intermediate-lived nuclei taken relative to stable or long-lived nuclei, in effect, determine the rate of nucleosynthesis averaged over their lifetime relative to the total average $\langle p \rangle$. Thus a very short-lived nucleus $\lambda T \gg 1$ only gives $p(T)/\langle p \rangle$ whereas a longer-lived nucleus gives the average $\overline{p(T)}/\langle p \rangle$ where $\overline{p(T)}$ is the value of $p(\tau)$ near T averaged over the lifetime of the nucleus considered. Comparing $\overline{p(T)}$ with $p(T)$ tells how $p(\tau)$ is changing with time near T. In other words it gives the derivative of $p(\tau)$ at T.

5. Input Data

As mentioned above the standard long-lived chronometer pair used is ^{232}Th/^{238}U; the intermediate pair is ^{235}U/^{238}U; and the short-lived pairs are ^{129}I/^{127}I and ^{244}Pu/^{232}Th. ^{244}Pu and ^{129}I are not measured directly since they are extinct (^{244}Pu has been found terrestrially but in extremely small amount, Hoffman et al., 1971). Their presence at the formation of the solar system is implied by their daughter products ^{129}Xe and a peculiar mass spectrum of the heavy Xenon isotopes, which has been found to be due to ^{244}Pu fission ($\lambda^{\text{fission}}_{244}/\lambda^{\alpha}_{244} = 1.2 \times 10^{-3}$). The confirmation (Alexander et al., 1971) that this peculiar Xe spectrum was due to ^{244}Pu is the most significant recent development in nucleochronologies.

The input data for these pairs is shown in Table I. Although the equations simplify better when ^{235}U is taken relative to ^{232}Th, that requires knowledge of the relative elemental abundances, which are much more uncertain than isotopic abundances, thus the data is given for ^{235}U/^{237}U.

The data given in Table I are similar to that used by Schramm and Wasserburg (1970), however there are some important differences.

One of the most significant is that the Podosek (1972) value 0.015 for $(^{244}$Pu/^{238}U)$_{T+\varDelta}$ is used as the standard rather than the Wasserburg et al. (1969)

TABLE I

Input data

Age of solid bodies in the solar system $t = 4.6 \pm 0.1 \times 10^9$ yr.	Reference Wasserburg and Burnett (1969)

Decay Constants (in yr^{-1})

$\lambda_{232} = 4.99 \times 10^{-11}$

$\lambda_{238} = 1.537 \times 10^{-10}$

$\lambda_{235} = 9.72 \times 10^{-10}$

$\lambda_{244} = 8.474 \times 10^{-9}$

$\lambda_{129} = 4.077 \times 10^{-8}$

Relative Abundances

$(^{129}I/^{127}I)_{T+\Delta} = 1.07 \pm 0.04 \times 10^{-4}$ — Hohenberg (1969)

$(^{244}Pu/^{232}Th)_{T+\Delta} = 0.0062^{+0.008}_{-0.002}$ — Podosek (1972) $(^{244}Pu/^{238}U \simeq 0.015)$; uncertainty includes Wasserburg *et al.* (1969a)

$(^{235}U/^{238}U)_{now} = \dfrac{1}{137.8};$ — U.S. Atomic Energy Commission Report

$\quad (^{235}U/^{238}U)_{T+\Delta} = 0.313 \pm 0.026$

$(^{232}Th/^{238}U)_{now} = 3.9^{+0.5}_{-0.9};$ — See Appendix Schramm and Wasserburg (1971)

$\quad (^{232}Th/^{238}U)_{T+\Delta} = 2.4(+0.35-0.15)$

r-process production ratios — Comments

$P_{129}/P_{127} = 1.5^{+1.4}_{-0.5}$ — Fowler (1972); range includes Seeger *et al.* (1965)

$P_{244}/P_{232} = 0.47 \pm 0.1$ — Uniform, range based on Seeger and Schramm (1970)

$P_{235}/P_{238} = 1.5 \pm 0.5$ — Uniform with 20 % odd-even effect; large uncertainty since odd-even effect could be as large as 40 % (Blake and Schramm, 1972)

$P_{232}/P_{238} = 1.9^{+0.2}_{-0.3}$ — Uniform, range based on Seeger and Schramm (1970)

$R(ij) = (p_i/p_j)/(N_i/N_j)_{T+\Delta}$

$R(129, 127) = (1.4^{+1.4}_{-0.5}) \times 10^4$

$R(244, 232) = 76^{+60}_{-50}$

$R(235, 238) = 4.8^{+2.2}_{-1.8}$

$R(232, 238) = 0.79^{+0.14}_{-0.21}$

value of 0.035. The Podosek (1972) value is based on his re-analysis of the Podesok (1970) data from a whole rock sample of the meteorite St. Severin. The Wasserburg et al. (1969a) value is the $^{248}Pu/^{237}U$ value for the whitlockite mineral separate from the St. Severin meteorite. Wasserburg et al. (1969a) also measured the whole rock sample and obtained a result similar to Podosek (1970). Since these numbers are based on only one meteorite, it is of course uncertain as to what the true $^{244}Pu/^{232}Th$ value is. It is felt here that an average whole rock sample might be less subject to chemical fractionation effects than a particular mineral. However, this is definitely an open question and can only be resolved by ^{244}Pu measurements in other meteorites, and by studies of Pu-U and Pu-Th chemical fractionation. Recently, Crozaz et al. (1972) investigated U-Th distributions in meteorites and found large amounts of fractionation occurred. Thus, it is probable that Pu would also be fractionated relative to U and Th which makes it more difficult to rely on the results from a single meteorite. In addition, Crozaz et al. found the U and Th (and thus probably the Pu) was almost exclusively in phosphate grains (whitlockite, chlorapatite etc.). If all the Pu and U in the meteorite were in phosphate grains, then the very small phosphate grains might have lost their fission Xe whereas the larger grains of the whitlockite mineral separate might not. Thus, it may be that the St. Severin whitlockite gives a more representative sample than the whole meteorite. For this reason, the uncertainty on the $^{244}Pu/^{232}Th$ ratio in Table I is taken to be large enough to include the whitlockite results.

In addition to the ^{244}Pu abundance, the current data use slightly different relative production rates than Schramm and Wasserburg. From examining the empirical r-process abundances (Seeger et al., 1965; Cameron, 1968) it can be seen that the iodine isotopes lie on the low A side of the $A \sim 130$ r-process peak; thus, $P_{129}/P_{127} \geqslant 1$. From looking at Te isotopic ratios nearby Fowler (1972) estimated $P_{129}/P_{127} = 1.5$. However, it is possible that the ratio P_{129}/P_{127} could be as high as 3 (Seeger et al., 1965).

The Actinide r-process ratios are based on the number of α-decaying progenitors to each long-lived nucleus (Burbidge et al., 1957). By using different mass formulae in an r-process calculation, it is possible to populate these progenitors in various ways. However, Seeger and Schramm (1970) showed that, on the average, the different progenitors are approximately uniformly populated; thus the ratio of the number of progenitors should approximately yield the r-process production ratios. Some workers (e.g. Fowler and Hoyle, 1960) have tried to use the rare earths to indicate the relative abundance of the progenitors and to indicate odd-even effects. However, Schramm and Fowler (1971) showed that the rare earth abundances might be due to fission rather than a regular r-process; thus they would not imply anything about relative actinide production rates. There is the possibility of an odd-even effect due to the large neutron-induced-fission cross section of the odd-A actinides. Blake and Schramm (1972) have shown that it is possible for the r-process nuclei to have been exposed to enough post-r-process neutrons to reduce the P_{235}/P_{238} ratio by 40%. However a more probable value would be a 20% reduction. It should be noted that the standard ratio P_{232}/P_{238} is much greater than the standard ratios used by Schramm and Wasserburg (1970) or Fowler (1972). This increase yields a resulting chronology with a much shorter time scale, similar to that found by Dicke (1969) when a different N_{232}/N_{238} was used.

6. Model Independent Results

From the value of $R(232, 238)$ it can be seen that

$$\Delta^{max}_{232, 238} \simeq 2.20 \times 10^9 \text{yr}.$$

However, values as large as 5.3×10^9 cannot be ruled out. This indicates that the mean age of the r-process elements $t + \Delta^{max}_{max} \approx 6.8 \times 10^9$ yr with an upper limit of $\approx 10^{10}$ yr. For most models $\Delta^{max} \lesssim T \leqslant 2\Delta^{max}$; thus the duration of r-process nucleosynthesis is $\lesssim t + 2\Delta^{max}$. It is currently thought that the r-process occurs in exploding massive stars, $M \gtrsim 4 M_\odot$. Since massive stars evolve quite rapidly, this means that the duration of r-process nucleosynthesis is approximately the age of the galaxy T_G. Thus

$$T_G \leqslant t + 2\,\Delta^{max} \leqslant 15 \times 10^9 \text{yr}.$$

This can be compared with the current value of Hubble's constant $H_0 = 55$ km/sec/mpc (Sandage, 1972), which yields an implied age for the universe, T_U, of $11 \times 10^9 = $ $= 2/3 H_0 \lesssim T_U \lesssim 1/H_0 = 17 \times 10^9$ where the range comes from the uncertainty about the deceleration parameter q_0. Current estimates, taking into account galactic evolution, put $q_0 \ll \frac{1}{2}$ (Tinsley, 1972; Sandage, 1972) which is consistent with the implications of D/H (see for example Reeves et al., 1972) which also imply a low value for q_0. Low values of q_0 correspond to higher values for T_U. Thus the age of the universe is quite comparable to the upper limit on the age of the galaxy obtained from completely independent means. The age of globular clusters has been estimated as $\sim 13 \pm 3 \times 10^9$ yr (Rood, 1972), which also yields a quite consistent picture. However, it should be kept in mind that the standard $R(232, 238)$ value yields a time scale which is somewhat less than the implied age of the universe.

From the short-lived chronometers it is possible to estimate

$$10^8 \lesssim \Delta \lesssim 2 \times 10^8 \text{yr}.$$

This time scale comes about primarily from the ^{129}I constraint. The standard parameter values yield $T^* \sim \Delta^{max}_{232, 238}$ which implies that $p(T) \sim \langle p \rangle$, with at most a small increase of $p(T)$ above $\langle p \rangle$ at T. Similarly the ^{235}U constraint shows $p'(T)$ to be small but slightly positive. It is of course possible to make a small parameter change well within the uncertainties and get $p'(T) = 0$ and $p(T) = \langle p \rangle$. Whether or not there is a bump or spike in $p(\tau)$ at T is quite sensitive to ^{244}Pu. High values of $(^{244}\text{Pu}/^{232}\text{Th})_{T+\Delta}$ [low values of $R(244, 232)$] such as the St. Severin whitlockite value require a spike at the end (Wasserburg et al., 1969b; Hohenberg, 1969; Fowler, 1972; Kohman, 1972) whereas lower values near the St. Severin whole rock value do not require any increase in $p(\tau)$ over the average rate at T.

7. Interpretation of Models

In order for a model to be valid, requires it to agree with the model-independent result. Thus, model-independently, the approximate time scales have been determined. However, within the parameter uncertainties, it is possible to use many different models. For example, Truran and Cameron (1971), Talbot and Arnett (1972) and Reeves (1972) have all discussed different models for the chemical evolution of the Galaxy, and all are consistent with nucleo-cosmochronology, to within the uncertainty in the parameters.

One thing a model does is give a physical interpretation to the model-independent parameters. Of course, total duration T and the mean age $T-\langle\tau\rangle$ are readily interpretable in any model. The distribution $p(\tau)$ is usually taken as the distribution of supernovae with time. With most models assuming rapid enough mixing of the supernovae ejecta in the galaxy that $p(\tau)$ is a continuous rather than discrete function. The point where different models give somewhat different interpretations is with regard to a possible final spike and $\varDelta$. If a spike is needed, it is not clear whether it is a single supernovae contributing an extra amount to the solar system due to it's proximity, or if it is a mixture of many supernovae mixing together in the cloud out of which the solar system eventually formed (Reeves, 1972). Whether or not a spike at T existed, there is still the question of the interpretation of $\varDelta$. In the work of Wasserburg *et al.* (1969b) it was thought (as mentioned by Cameron, 1962) that the spike was a single supernovae which triggered the collapse of the protosolar cloud to form the solar system, with $\varDelta$ being the time between solidification of solid bodies in the solar system and when the protosolar gas started its collapse and separated from the galactic gas. Recent work by Cameron (1972) and Reeves (1972) has shown that such a model runs into hydrodynamic difficulties. Once a collapse starts it is difficult to imagine why it should take $\sim 10^8$ yr before objects form; and then, why objects should suddenly solidify on a time scale of a few million years as determined by the $^{129}\text{I}-^{129}\text{Xe}$ isochronism and by the Rb-Sr techniques of Papanastassiou and Wasserburg (1971).

The current solar system model of Cameron (1972) has the solar system collapse and form in $\sim 10^7$ yr which means that the $\varDelta \sim 10^8$ yr. would be measuring something else. Reeves (1972) and Cameron (1972) noticed that $\varDelta$ is about the time scale of a galactic revolution, and thus proposed that $\varDelta$ might be related to the mixing time scale of the galaxy. In fact, if it is assumed that supernovae occur predominantly in spiral arms, then $\varDelta$ might plausibly represent the time between solar system solidification and when the protosolar gas was last in a spiral arm. This type of explanation would fit in well with the density wave theory of galactic structure (Lin and Shu, 1964).

In order to better understand the mathematical nature of model dependent results, the simple exponential model of Fowler (1972) will now be shown (see Figure 1b). In this model

$$p(\tau) = A \exp(-\tau/\theta) + B\delta(\tau - T)$$

Exponential Model

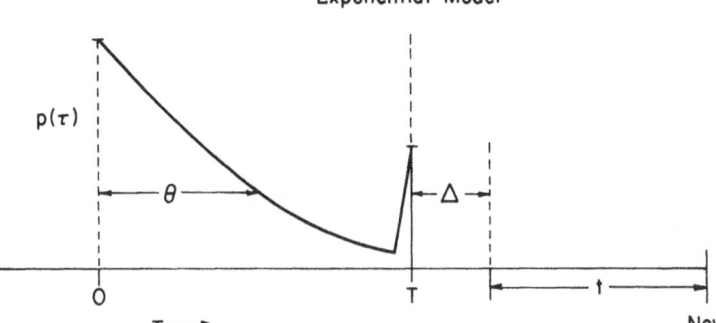

Fig. 1b.   The exponential model

$$p(\tau) = A \exp(-\tau/\theta) + B\delta(\tau - T)$$

Usually the normalization parameters A and B are combined so as to give a parameter C which represents the fraction of a stable nucleus synthesized in the final spike.

where θ is the time constant for the model and A and B are normalization constants. Since only relative abundances are important A and B can be combined to yield a single constant C which represents the fraction of stable material synthesized in the final event at time T. Thus the model has 4 parameters T, $\varDelta$, θ and C. (C and θ are assumed to be $\geqslant 0$), and with the 4 nucleochronometric pairs, it yields 4 simultaneous equations in 4 unknowns.

The exponential model is convenient to work with because it approximates star formation rates in a reasonable manner that is roughly consistent with the approaches of Truran and Cameron (1971) and Talbot and Arnett (1972). The model is also versatile in that it can yield solutions with or without a final spike by having C be non-zero or zero. It can also go to the limits of a continuous uniform model ($p(\tau) = \langle p \rangle$ for all τ; $0 < \tau < T$) or to a two spike model (with one spike at the beginning ($\tau = 0$) and another $\varDelta$ years before solidification). These limits are obtained by varying θ from ∞ to 0. In this model C is related to $p(T)/\langle p \rangle$, and the mean age of the elements prior to solar system solidification, $\varDelta^{max}$, can be used to estimate T, $\varDelta^{max} \leqslant T \leqslant 2\varDelta^{max}$. Where the longer values of T are obtained for large values of θ and the lower values for small values of θ.

Figure 2 shows how the parameter $\varDelta$ varies when each of the $R(ij)$'s is varied within its stated uncertainty while the other $R(ij)$'s are held at their standard values. The largest changes in $\varDelta$ result from the variation of $R(129, 127)$ and $R(244, 232)$. It should be noted that $\varDelta$ is rather insensitive to parameter changes and is always between ~ 1 and 2×10^8 yr.

Figure 3 is similar to Figure 1, but in this figure the variations of C are shown. Notice that the relative size of the spike has its maximum range of variation from $R(244, 232)$. It can be seen that within the uncertainties, values for C as high as $\sim 8\%$ are possible as are extremely small values of $\sim 1\%$. By slightly varying more than one $R(ij)$ from its standard value, it is possible to have solutions with $C=0$.

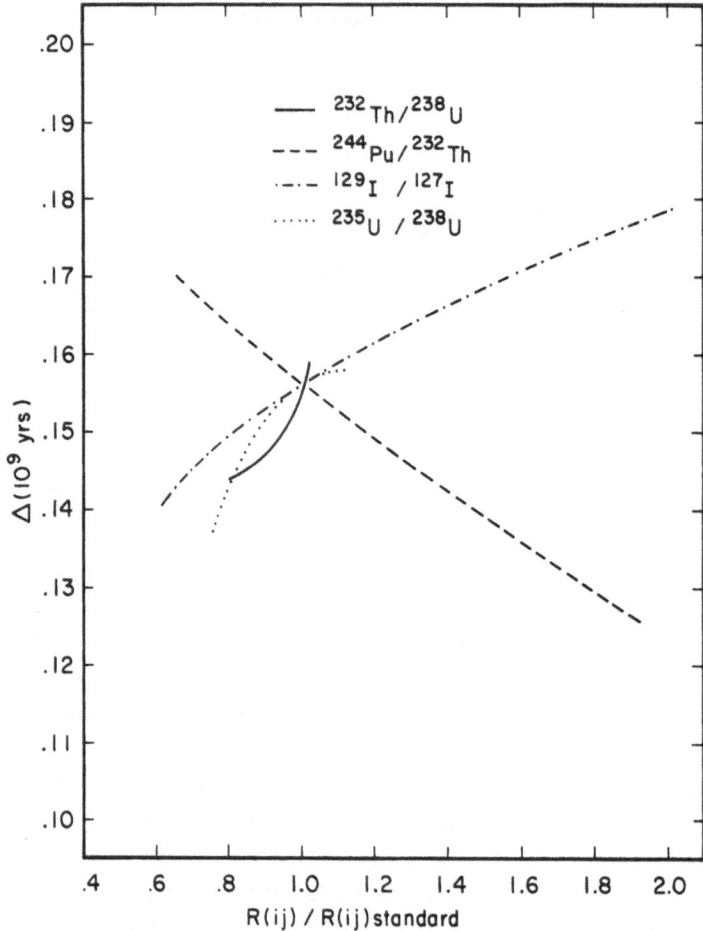

Fig. 2.   The variation of Δ in the Fowler (1972) exponential model when each $R(ij)$ is varied within its range of uncertainty, holding the remaining $R(ij)$'s at their standard values. Notice that there is very little variation of Δ; $1 \times 10^8 \lesssim \Delta \lesssim 2 \times 10^8$.

Figure 4 shows the variations of T. Notice that these variations are most sensitive to $R(232, 238)$, and that T is virtually independent of the short-lived chronometers ^{129}I and ^{244}Pu, as would be expected.

Since T varies from 2 to over 10×10^9 yr, this means that the age of the galaxy T_G can vary from 7 to 15×10^9 yr in this model. This is quite consistent with our model-independent conclusions discussed previously. Figure 5 shows the variations of θ. Here it is seen that θ can vary from a uniform model, with $\theta = \infty$, to a spiked model with $\theta = 0$, depending on the relation between Th and the U isotopes. This relationship is hinged on the fact that ^{235}U needs $\sim 10^9$ yr to decay, thus for certain values of T and $R(235, 238)$ it is necessary to have the bulk of the nucleosynthesis early so that ^{235}U can decay. θ is virtually independent of ^{129}I and ^{244}Pu for most parameter values. However when $R(244, 232)$ is very low, so that a

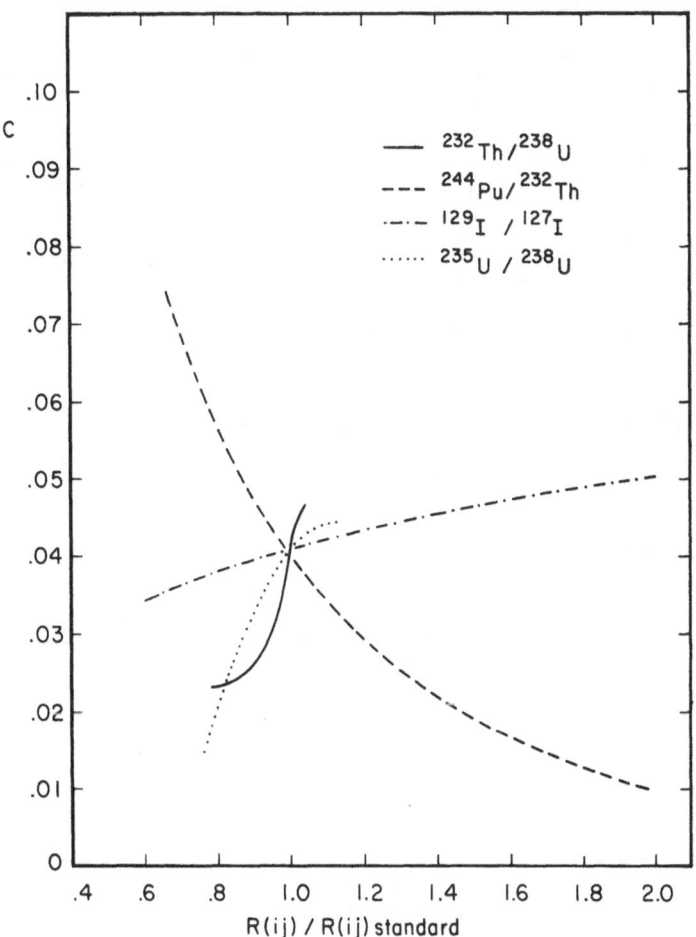

Fig. 3. Similar to Fig. 2 but showing the variation of C, the fraction of a stable nucleus synthesized in the final spike. Notice that C can vary from $< 1\%$ to $> 7\%$ depending on values of $R(244,232)$.

large spike is required, then θ must be small, so that the ^{235}U synthesized early has time to decay before the ^{235}U made in the spike is added.

It can be seen that the model-independent conclusions are quite valid. It can also be seen that until the uncertainties on the parameters are narrowed it will be difficult to determine a detailed model for the galaxy from nucleochronology alone.

8. Future Possibilities

In the future, one might hope for improvements in the determination of the value of $(^{244}$Pu$/^{232}$Th$)$ at Xe-retention. Such improvement could result from more meteorites being studied, as well as a better understanding of Pu fractionation. With the possible exception of ^{232}Th$/^{238}$U the other abundance determinations are fairly well

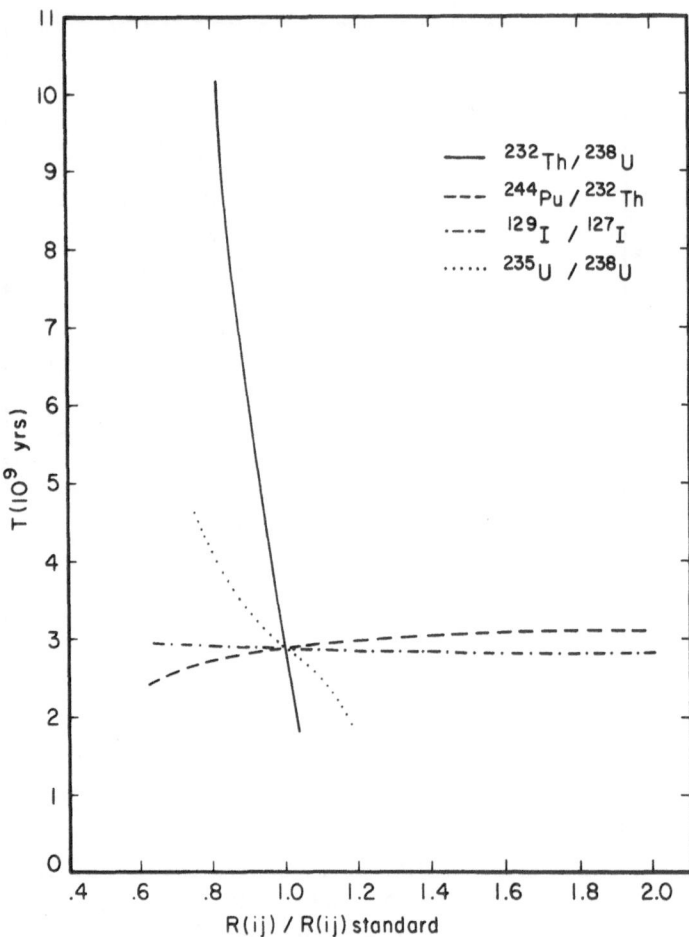

Fig. 4.   Similar to Figure 2 and Figure 3, showing variation T. Notice the strong sensitivity of T to $R(232,238)$ and the virtual independence of T to changes in $R(244,232)$ and $R(129,127)$

established. The r-process production estimates can probably not be improved to any great degree until a more accurate method for estimating nuclear masses far from the valley of β stability is developed by nuclear theorists. Since physicists have been working on nuclear mass formulae for over 30 years, this possibility may be long in coming. There is also the additional problem of understanding the true nature of the r-process. The one production ratio which might become better known is P_{129}/P_{127}, since this conceivably could be determined from ^{129}Xe and ^{127}I abundances.

For the r-process, our best hope for the future lies with ^{187}Re (Clayton, 1964) $(\tau_{1/2} \simeq 4.3 \times 10^{10}$ yr), since this chronometer would be capable of giving the mean age directly, with only negligible model dependent corrections.

$$\Delta_{187}^{max} = \frac{1}{\lambda_{187}} \ln\left[1 + \frac{(^{187}\text{Os})_c}{(^{187}\text{Re})}\right],$$

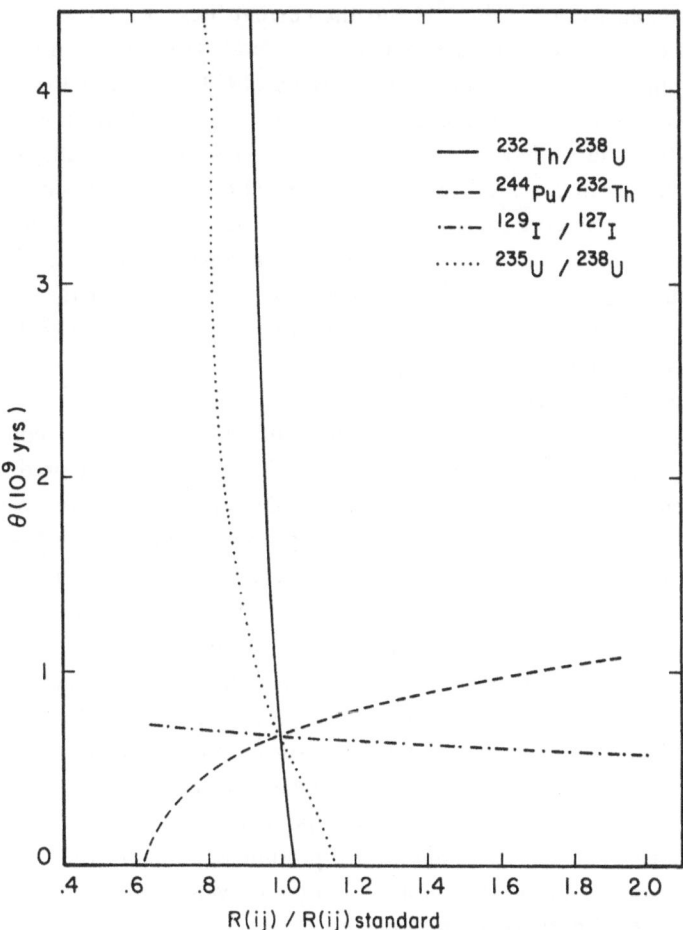

Fig. 5. Similar to the previous 3 figures showing the variation of the parameter θ in the exponential model. $\theta = \infty$ is the uniform case with a spike of strength C at $\tau = T$, whereas $\theta = 0$ is a 2 spike case with events at $\tau = 0$ and $\tau = T$.

where

$$\frac{(^{187}\mathrm{Os})_c}{(^{187}\mathrm{Re})} \simeq 0.23\left[1 - \frac{1.20\sigma_{186}}{\sigma_{187}}\right]$$

with $\sigma_{186}/\sigma_{187}$ being the ratio of the $^{186}\mathrm{Os}$ and $^{178}\mathrm{Os}$ neutron capture cross sections. This chronometer is all the more promising, because all parameters entering into the determination of the mean age $\Delta_{187}^{\max}$ are experimentally measurable. Unfortunately $\sigma_{186}/\sigma_{187}$ has not been measured yet, and λ_{187} has large uncertainties in it's determination. Using a best guess for $\sigma_{186}/\sigma_{187}$ of 0.4 ± 0.1 and taking into account the full range of uncertainty in λ_{187} yields

$$4 \times 10^9 \lesssim \Delta_{187}^{\max} \lesssim 11 \times 10^9.$$

since $\Delta^{\max} < T < 2\Delta^{\max}$. This implies $9 \times 10^9 \,\mathrm{yr} < T_G < 27 \times 10^9 \,\mathrm{yr}$, which overlaps the range determined from ^{232}Th/^{238}U but the current 'best' values are off by several 10^9 yr. Since cross section estimates may be considerably off, there is no need to doubt the ^{232}Th number at the present time. Clayton (1969) proposed that possibly the large discrepancy between Re and Th chronologies was caused by an enhanced decay rate for ^{187}Re in stars. Perrone and Clayton (1972) have calculated the magnitude of such an enhancement, and Talbot and Arnett (1972) has examined what effect it might have on ^{178}Re chronologies and has found it to be small, $\lesssim 10\%$. Thus, the importance of measuring the Os cross sections is again brought to the forefront.

The possibility of developing chronometers for processes other than the r-process is now becoming possible. In particular Audouze and Schramm (1972) have shown that the p-process nucleus ^{146}Sm ($\tau_{1/2} = 1.2 \times 10^8$ yr) will be produced in sufficient amount so that anomalies in the isotopic abundance of the daughter nucleus, ^{142}Nd, should be detectable. Such anomalities will occur when the isotopic composition of Nd in meteoritic material having different Nd/Sm ratios is compared. ^{146}Sm is a short-lived chronometer, which would tell us if the p-process was also occurring just before the solar system formed. A possible s-process chronometer (Audouze *et al.*, 1972) is ^{176}Lu ($\tau_{1/2} \simeq 2.2 \times 10^{10}$ yr), however effects from it are too small to be detectable with present techniques. However, it is possible to use this chronometer to predict certain s-process neutron capture cross sections, and the branching ratio for neutron capture to the short and long-lived isomers of ^{176}Lu. Another possible s-process chronometer is ^{205}Pb ($\tau_{1/2} \simeq 3 \times 10^7$ yr). Attempts at developing it have been made by Kohman and Huey (1972). If a short-lived s-process chronometer were developed, then it may be possible to check the galactic density wave explanation of Δ. s-process sources are presumably red giants, which populate both the interarm and arm regions, whereas the r-process sources are presumably supernovae which seem to occur mostly in the arms. Therefore, a different value for Δ should result for s- versus r-process short-lived chronometers.

9. Conclusion

Nucleochronometers are capable of obtaining model independent results about the history of the galaxy. However, detailed conclusions about galactic evolution cannot at present be given, because of the large uncertainties in the input parameters. Future measurements, particularly on the ^{244}Pu problem and the development of the ^{187}Re chronology, could enable significant improvement in the state of r-process chronology. The development of ^{146}Sm as a p-process chronometer would also be an important breakthrough in the field, as would the development of any other chronometer.

Acknowledgements

I would like to thank D. Burnett, A. G. W. Cameron, F. Podosek and H. Reeves for interesting discussions on various aspects of the work. I have also had the pleasure of having my opinions on this subject molded by W. A. Fowler and G. J. Wasserburg.

Work supported in part by the National Science Foundation. [GP-27304 and GP-28027 at Cal Tech and GP-32051 at Univ. of Texas].

References

Alexander, E. C., Lewis, R. S., Reynolds, J. H., and Michel, M. C.: 1971, *Science* **172**, 837.

Audouze, J. and Schramm, D. N.: 1972, *Nature* **237**, 447.

Audouze, J., Fowler, W. A., and Schramm, D. N.: 1972, *Nature Phys. Sci.* **238**, 8.

Blake, J. B. and Schramm, D. N.: 1972, *Astrophys. J.*, in press.

Burbidge, E. M., Burbidge, G., Fowler, W. A., and Hoyle F.: 1957, *Rev. Mod. Phys.* **29**, 547.

Cameron, A. G. W.: 1962, *Icarus* **1**, 13.

Cameron, A. G. W.: 1968, in L. H. Ahrens (ed.), *Origin and Distribution of the Elements*, Pergamon Press, New York, p. 125.

Cameron, A. G. W.: 1972, Preprint and private communication.

Clayton, D. D.: 1964a, *Astrophys. J.* **139**, 637.

Clayton, D. D.: 1964b, *Nature* **224**, 56.

Crozaz, G., Burnett, D. S., and Walker R.: 1972, *Proc. Cosmochemistry Symposium*, Cambridge, Mass.

Dicke, R. H.: 1969, *Astrophys. J.* **155**, 123.

Fowler, W. A.: 1972, in F. Reines (ed.), *Cosmology Fission and Other Matters. A Memorial to George Gamow*, Colorado Assoc. Univ. Press, Boulder.

Fowler, W. A. and Hoyle F.: 1960, *Ann. Phys.* **10**, 280.

Hoffman, D. C., Lawrence, F. O., Mewherter, J. L., and Rourke, F. M.: 1971, *Nature* **234**, 132.

Hohenberg, C. M.: 1969, *Science* **166**, 212.

Hohenberg, C. M., Podosek, F. A., and Reynolds, J.: 1967, *Science* **156** 202.

Kohman, T. and Huey, X. X.: 1972, *Proc. Cosmochemistry Symposium*, Cambridge, Mass.

Lin, C. C. and Shu, F. H.: 1964, *Astrophys. J.* **140**, 646.

Papanastassiou, D. A. and Wasserburg, G. J.: 1971, *Earth Planetary Sci. Letters* **11**, 37.

Perrone, F. and Clayton, D. D.: 1972, to be published.

Podosek, F.: 1970, *Earth Planetary Sci. Letters* **8**, 183.

Podosek, F. A.: 1972, *Geochim. Cosmochim. Acta.* **36**, 755.

Reeves, H.: 1972, *Astron. Astrophys.* **19**, 215.

Reeves, H., Audouze, J., Fowler, W. A., and Schramm, D. N.: 1972. *Astrophys. J.*, in press.

Rood, R.: 1972, private communication.

Schramm, D. N. and Fowler, W. A.: 1971, *Nature* **231**, 103.

Schramm, D. N. and Wasserburg, G. J.: 1971, *Astrophys. J.* **162**, 57.

Seeger, P. A. and Schramm, D. N.: 1970, *Astrophys. J.* **160**, L157.

Seeger, P. A., Fowler, W. A., and Clayton, D. D.: 1965, *Astrophys. J. Suppl.* **11**, 121.

Talbot, R. J. and Arnett, W. D.: 1972, to be published.

Tinsley, B.: 1972, private communication.

Truran, J. W. and Cameron, A. G. W.: 1971, *Astrophys. Space Sci.* **14**, 179.

Wasserburg, G. J. and Burnett, D. S.: 1967, *Earth Planetary Sci. Letters* **2**, 397.

Wasserburg, G. J., Huneke, J. C., and Burnett, D. S.: 1969a, *J. Geophys. Res.* **74**, 4221.

Wasserburg, G. J., Schramm, D. N., and Huneke, J. C.: 1969b, *Astrophys. J.* **157**, L91.

A COSMOCHEMICAL VIEW OF COSMIC RAYS
AND SOLAR PARTICLES

P. B. PRICE*

Dept. of Physics, University of California, Berkeley 94720, Calif., U.S.A.

Abstract. The composition of cosmic rays and solar particles is reviewed with emphasis on the question of whether they are representative samples of Galactic and solar matter. The composition of solar particles changes with energy and from flare to flare. A strong excess of heavy elements at energies below a few MeV/nuc decreases with energy, and at energies above ~ 15 MeV/nuc the composition of solar particles resembles that of galactic cosmic rays somewhat better than that of the solar atmosphere. The elements Ne through Pb have remarkably similar abundances in cosmic ray sources and in the matter of the solar system. The lighter elements are depleted in cosmic rays, whereas U and Th may be enriched or not, depending on whether the meteoritic or solar abundance of Th is used. Two prototype sources of cosmic rays are considered: gas with solar system composition but enriched in elements with $Z \cdot 8$ during acceleration and emission (by analogy with solar particle emission), and highly evolved matter enriched in r-process elements such as U, Th and transuranic elements. The energy-dependence of cosmic ray composition suggests that both sources may contribute at different energies.

1. General Remarks

At a Cosmochemistry Symposium it is appropriate to ask what it is that cosmic rays sample and whether the information they provide is significantly different from stellar spectroscopic data. Because they are electrically charged, in traversing galactic magnetic fields their directions are randomized and we should consider the possibility that their composition might represent a better average of the elemental composition of our Galaxy than do the meteorites or the solar atmosphere or individual stellar abundances. But because cosmic rays are highly nonthermal in nature, having energies that range from as low as a few tens of MeV/nucleon up to as high as $\sim 10^{15}$ MeV, we must be wary of possible strongly selective effects both in the raw material from which they are made and in the stages by which they are accelerated and propagated through space. In that case we might be observing high energy matter from the envelope of a neutron star, or from the atmosphere of a white dwarf, or from the debris of a supernova enriched in the products of explosive nucleosynthesis.

Several steps are necessary in order to build up a reasonably accurate picture of the source material from which cosmic rays are made:

1. Detectors located at the top of, or beyond, the Earth's atmosphere must be large enough to collect good statistics and must have adequate resolution to establish actual values, not just upper limits, for the rarer odd-Z nuclei as well as the more abundant even-Z nuclei.

2. A model of cosmic ray propagation must take account of modulation and adiabatic deceleration within the solar system resulting from interactions with expand-

---

* Miller Institute Professor, 1972–73.

A. G. W. Cameron (ed.), Cosmochemistry, 69–88. All Rights Reserved

ing solar magnetic fields; and it must distribute the sources spatially and temporally so that those elements certain to be absent in the sources are produced in the right amounts to match observations at earth. Values of cross-sections as a function of energy must be known or calculated for all the nuclear transmutations that take place during propagation.

3. Following these steps, we can prepare a table of abundances of cosmic rays before propagation from the sources but after injection and initial acceleration out of source regions.

4. Finally, corrections must be made for selection effects at the sources that may result in the systematic depletion of some elements and the enrichment of others. It is only recently that enough progress has been made to justify investigating the last step. This is where the study of energetic particles emitted in solar flares is extremely useful and forms an integral part of a talk on galactic cosmic rays. With solar particles the source is known, spallation is negligible, and propagation is not likely to introduce serious complications. To the extent to which we can believe solar spectroscopic data, we can study selection effects for a known cosmic accelerator – the sun – and try to draw analogies between solar particles and extra-solar particles. Thus, the final step involves the detailed study of solar particle abundances over a wide range of energies and averaged over a spectrum of flares of different strengths.

Our discussion of cosmic ray composition is facilitated by the fact that detailed reviews were presented as rapporteur talks at the Twelfth International Cosmic Ray Conference held in Hobart, Tasmania (Shapiro, 1971; Price, 1971; Simpson, 1971). A number of new developments in solar particle studies have occurred since the previous review paper on that subject some years ago (Fichtel and McDonald, 1967), and the present talk will have to serve as a guide to a very fluid situation.

Devices for identifying cosmic rays measure their rate of ionization (or energy deposition), which varies as $\sim Z^2/\beta^2$, and determine their velocity, β, separately from their radius of curvature in a magnetic field or from their total energy or range, or from their Cerenkov radiation. Active electronic detectors range in size from semi-conductor crystals ~ 1 cm^2 in size on satellites up to scintillators or spark chambers nearly 1 m^2 in size that have been flown on balloons or are being designed for the HEAO satellites due to be launched toward the end of this decade. Passive detectors made of layers of track-recording plastics or nuclear emulsions with enormous collecting area (~ 60 m^2 in one experiment) are used to study the very rare but astrophysically important cosmic rays near the end of the Periodic Table.

2. Cosmic Ray Composition

Figure 1 gives an overall picture of cosmic ray abundances at the top of the Earth's atmosphere, compared with solar system abundances. The immediate impression is of a remarkable similarity in the two distributions. One has to look closely to note the differences, which are of two types: (1) the presence of very rare nuclei suchs a Li, Be, B, F and the elements Cl to Mn, which are produced by spallation reactions

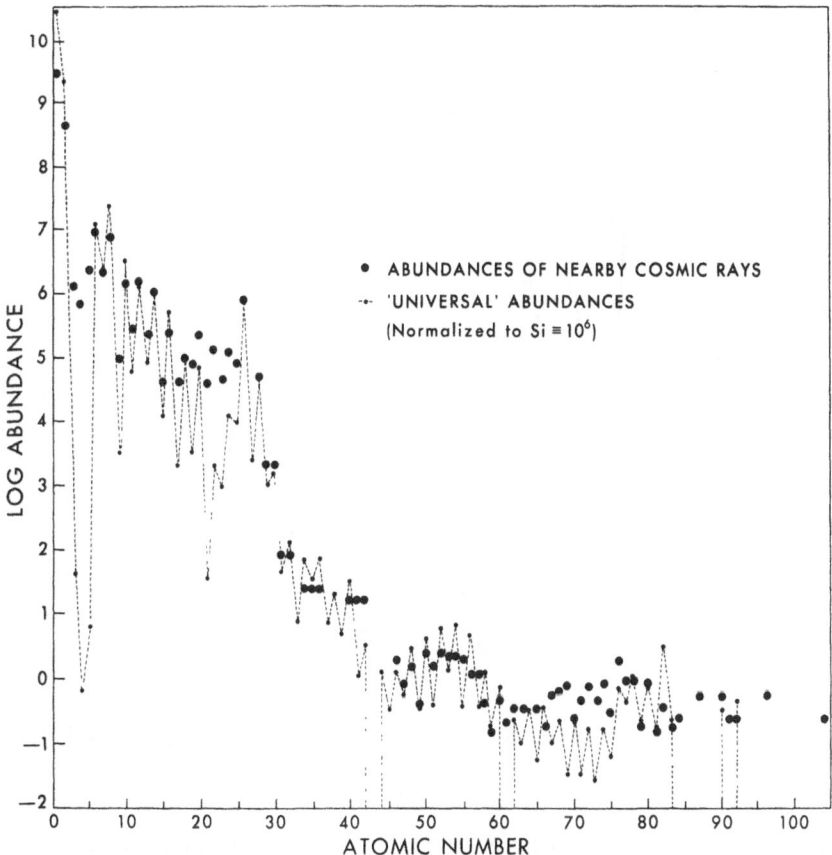

Fig. 1.   Composition of cosmic rays at 1 AU compared with solar system abundances. The recent photospheric values of Th and U (Grevesse, 1969), corrected for radioactive decay during the lifetime of the Sun, were used rather than the far lower meteoritic abundances of Th and U (Cameron, 1968).

during transport through the interstellar medium; and (2) depletions of no more than one order of magnitude of some of the light primary cosmic rays such as H and He. Some idea of the progress that has been made can be gained when we realize that less than a decade ago none of the cosmic rays heavier than Ni had been discovered and the nuclei from H to Ni were lumped into six groups – H, He, L, M, H and VH – because of inadequate resolution of individual elements.

Figure 2 gives a closer look at the most extensively studied region, from He to Fe. Figure 3 covers the heavier elements, for which, because of their extreme scarcity, statistics are still quite poor (Shirk *et al.*, 1972). The points indicated by squares represent the abundances at balloon altitude predicted by a model that assumes solar system abundances at cosmic ray sources. Only eight events assigned to the actinide group, $Z \geqslant 90$, have been detected thus far in almost a dozen balloon flights carrying large area detectors. Three of these events may be trans-uranic, but in view of differences of charge assignment to the same events by two different types of detectors –

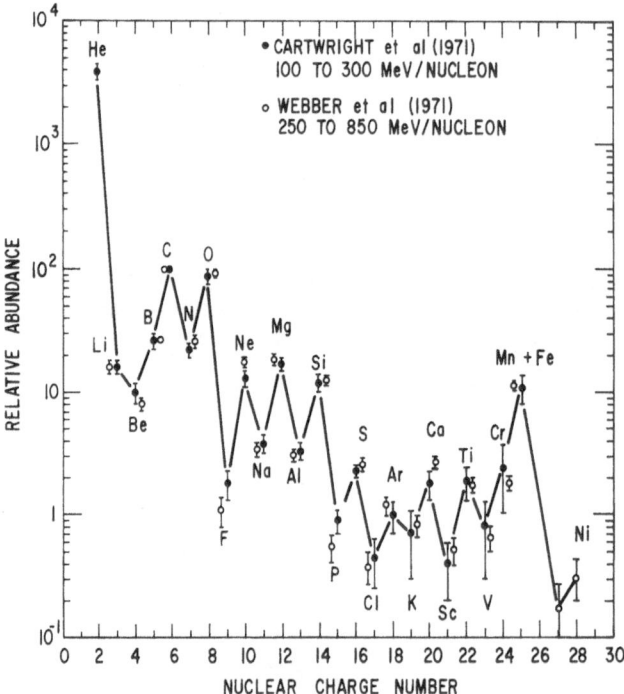

Fig. 2. Cosmic ray abundances at 1 AU for He to Ni measured at 100 to 850 MeV nuc^{-1} and normalized to carbon $\equiv 100$.

plastics and emulsions – one cannot rule out the possibility that they are U or Th (Price, 1971).

3. Energy Dependence of Composition

In the interval from He up to Fe, careful measurements with high-resolution detectors in satellites (Cartwright *et al.*, 1971), in balloons (Webber *et al.*, 1972; Casse *et al.*, 1971; Juliusson *et al.*, 1972; Ormes and Balasubrahmanyan, 1973; Webber *et al.*, 1973; Smith *et al.*, 1973), and on the lunar surface (O'Sullivan *et al.*, 1973; Chan and Price, 1973) have recently established that the composition of cosmic rays slowly changes with energy. The changes are of two types, as Figure 5 shows: the relative proportion of secondary nuclei such as the elements with $17 \leqslant Z \leqslant 25$ (Figure 5a) and the elements Li, Be and B decrease with increasing energy (Figure 5b). The changes are gradual and are hardly detectable at energies below a few GeV/nuc.

For the ultra-heavy cosmic rays the situation is quite different. An enrichment of cosmic rays with $Z > 60$ relative to Fe and lighter particles has been reported (Price, 1971; Shirk *et al.*, 1973) at the lowest energies studied – a few hundred MeV/nuc^{-1} – but not at energies above 1 GeV nuc^{-1}. The data, shown in Figure 6, provide encouragement to those groups designing experiments to search for trans-uranic cosmic rays at low energies with detectors on satellites.

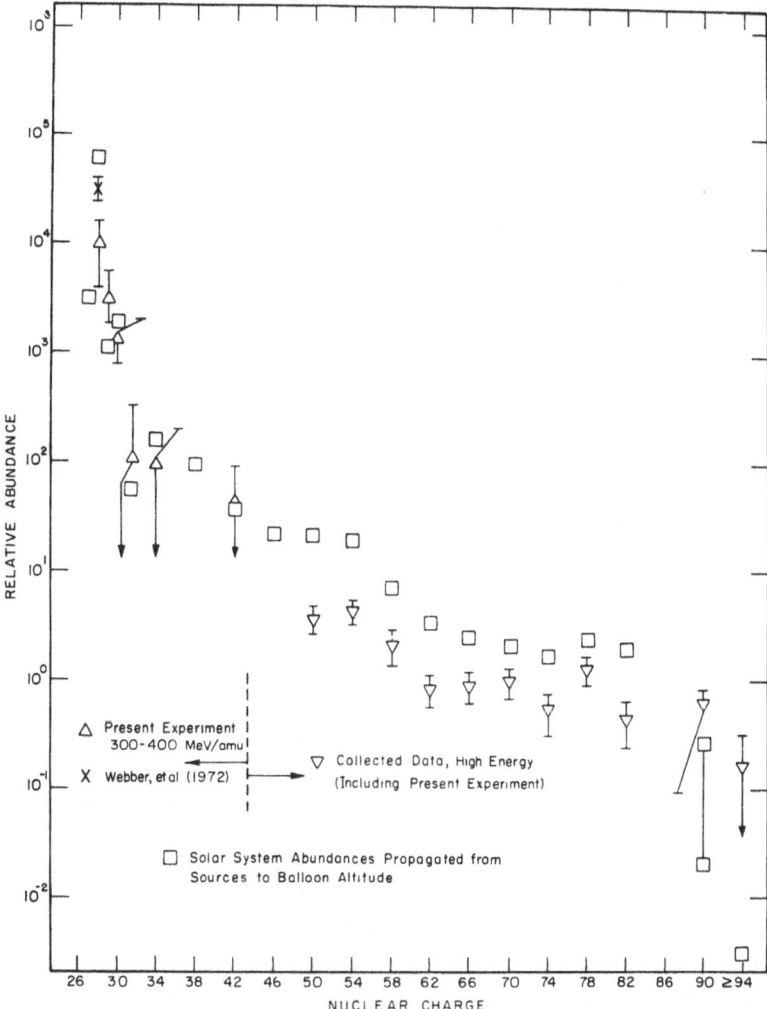

Fig. 3.   Abundances of cosmic rays with $Z \geq 26$ at a depth of ~ 3 g cm^{-2} in the atmosphere. In the region $26 \leq Z \leq 42$ the triangles gives fluxes of particles with an energy of 300 to 400 MeV nuc^{-1} that stopped in a stack of Lexan detectors (Shirk *et al.*, 1973). For $Z > 42$ the symbols Δ give data compiled from all sources (summarized by Shirk *et al.*, 1973) and are mainly for relativistic energies. The squares are the result of propagating solar system abundances (Cameron, 1968) through an exponential path length distribution in the Galaxy and down to balloon level. All data are normalized to $F_e \equiv 10^6$.

4. Time Dependence of Composition

Variations of flux on a time-scale of a few years have their origin in the eleven-year cycle of solar activity and tell us not about the cosmic radiation but about the helio-magnetosphere. To go back to astrophysically interesting time-scales we must exploit the 'fossil record'. Radiochemical studies of nuclides produced by cosmic ray interactions in meteorites are insensitive to possible short-term fluctuations in overall

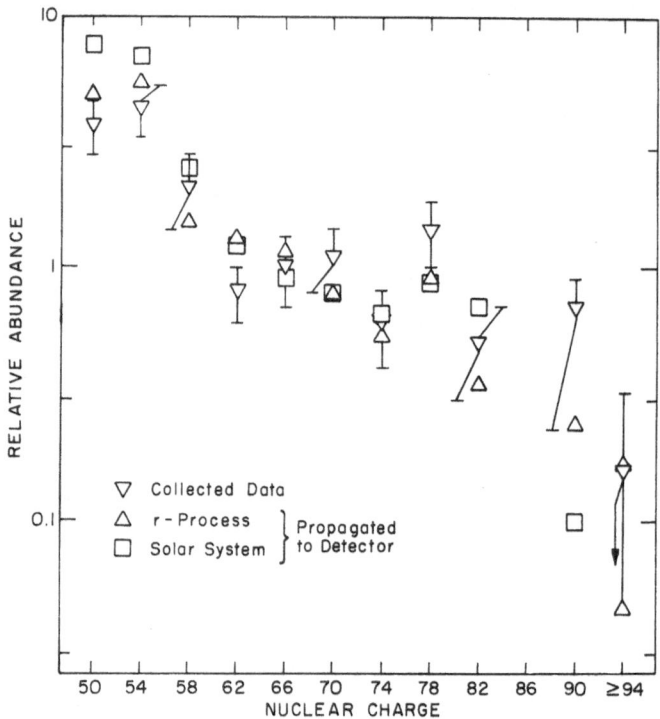

Fig. 4. The data are the same as in Figure 3, but trial source distributions with solar system composition (Cameron, 1968) and with r-process composition (Seeger *et al.*, 1965) were normalized to the data at $60 \leqslant Z \leqslant 83$ after propagation through an exponential path length distribution.

intensity but show that the long-term level has been constant to better than 50% over various time-intervals extending from the present back to about 10^9 yr (Arnold *et al.*, 1961). By counting tracks of heavy cosmic rays in meteorites of known cosmic ray exposure ages, it has been shown that the relative abundances of Fe and the trans-iron nuclei have not drastically changed over the last $\sim 10^7$ yr (Cantelaube *et al.*, 1967; Maurette *et al.*, 1968; Price *et al.*, 1971). A sufficiently careful study has not yet been made that one could rule out fluctuations of the kind predicted by Ramaty *et al.* (1970) for the extremely heavy nuclei such as U that have very short mean free paths for collisions with interstellar gas. They showed that time-variations of the order of a factor two in the fluxes of the heaviest cosmic rays would be expected if they come from point sources in space and time (e.g. supernova explosions).

5. Source Abundances of Cosmic Rays

Silberberg and Tsao (1973) have developed semi-empirical equations for calculating cross-sections for production of various daughter nuclides in collisions of various parent nuclides with interstellar gas. These equations are useful in supplementing the very meager cross-section data from accelerator bombardments of various targets.

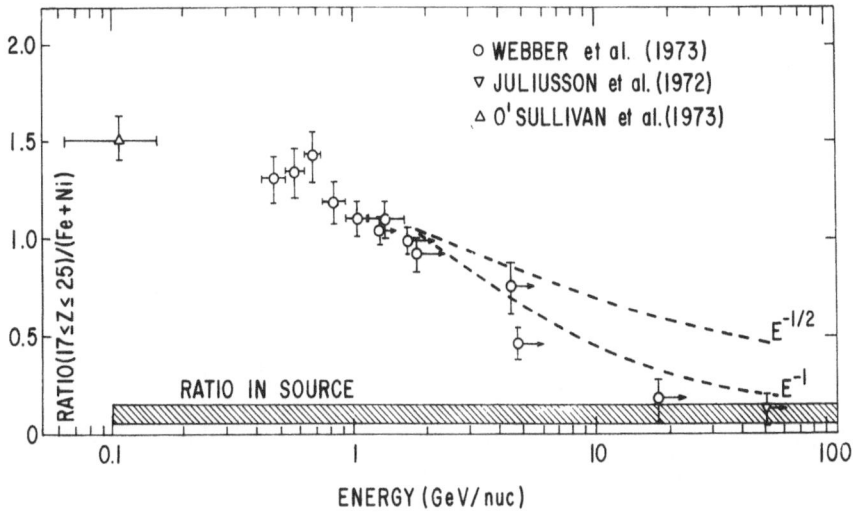

Fig. 5a. Energy dependence of cosmic ray composition: ratio of abundances of secondary to parent nuclei. The hatched lines refer to calculated ratios of source abundances (Table I).

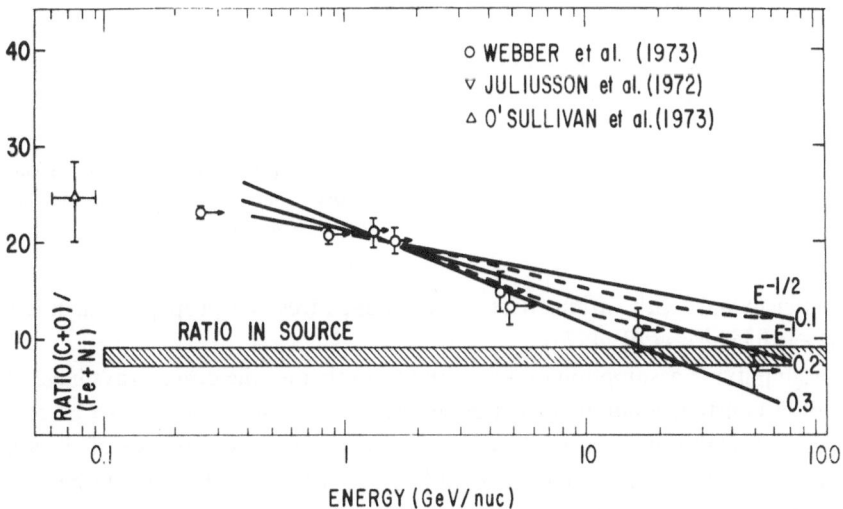

Fig. 5b. Energy dependence of cosmic ray composition: ratio of abundances of light to heavy primary nuclei. The hatched lines refer to calculated ratios of source abundances (Table I).

In a typical calculation of source composition the observed abundances are propagated backward through some distribution of path length until the fluxes of nuclei known to be extremely rare in the universe drop to zero. A pathlength distribution is regarded as acceptable if rare nuclei covering a very broad charge interval simultaneously disappear. The method has been discussed in detail in Shapiro and Silberberg (1970), by Webber *et al.* (1972), and more recently by Cartwright *et al.* (1973). For the high-Z cosmic rays, which have very short breakup mean free paths,

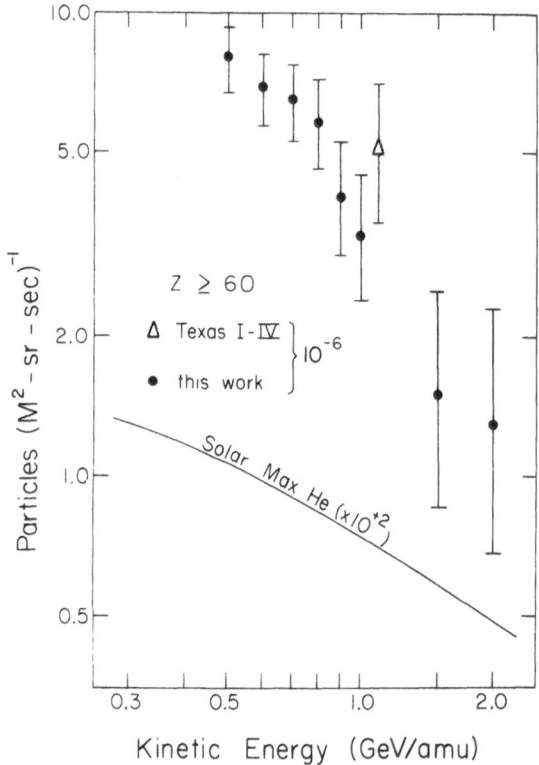

Fig. 6.   The integral energy spectrum for particles with $Z > 60$ (Shirk *et al.*, 1973) is considerably steeper than the spectra of He, CNO, and Fe and suggests that different sources may be contributing ultra-heavy cosmic rays of low energy.

it is better to start with various trial source abundances and propagate them forward to the earth (Shirk *et al.*, 1973).

The simplifying assumption of a few years ago that all the cosmic rays have passed through a constant amount of matter several g cm^{-2} thick has had to be replaced with a more elaborate picture. In order to account for the presence of ultra-heavy cosmic rays such as Pt, Pb, Th and U at levels at least as high as would be expected for solar system source abundances or for r-process source abundances (Figure 4), the pathlength distribution must be peaked at a very small value, considerably smaller than 0.8 g cm^{-1}, which is the mean pathlength for breakup of a U nucleus in a nuclear collision, but must include much larger values as well, in order to produce the observed secondary nuclei. Provided we consider only data at energies below a few GeV nuc^{-1}, an exponential distribution, $N(x) \sim \exp(-x/\lambda_0)$, with a leakage mean free path $\lambda_0 \approx 5$ g cm^{-2}, satisfactorily accounts for the observed abundances of the major secondaries that have been studied: H^2, He^3, LiBeB, F, $17 \leqslant Z \leqslant 25$, and the elements $60 < Z < 70$. If the cosmic rays are propagated and undergo their collisions in the Galactic disc where the density is ~ 1 atom cm^{-3}, the leakage lifetime corresponding to $\lambda_0 = 5$ g cm^{-2} would be ~ 3 m.y., whereas if most of their pathlength is traversed in a region of density $\sim 10^{-2}$ atom cm^{-3}, their lifetime would be ~ 300 m.y. So far it has not been technically possible to resolve radioactive isotopes such as ^{10}Be or ^{244}Pu that have suitable half-lives to provide direct information on leakage lifetimes of the cosmic rays.

One way to account for the gradual changes in composition shown in Figure 5 is to assume that λ_0 is a decreasing function of energy. Then at high energy the abundance ratios would approach those calculated for cosmic ray sources since very little matter would have been traversed and very little breakup would have occurred. We see in Figure 5 that the ratios at energies above ~ 20 GeV nuc^{-1} indeed seem to approach the calculated source ratios. Webber et al. (1973), Audouze and Cesarsky (1973), and Meneguzzi (1973) have simply postulated that λ_0 decreases with energy because high-energy cosmic rays leak out of the Galaxy more readily than do low-energy cosmic rays. In Figures 5a and 5b the dashed curves show how the ratios would change if λ_0 varied as $E^{-1/2}$ or as E^{-1}. The solid curves in Figure 5b show how the ratios would change if the exponents in the differential energy spectra of C and O are more negative by 0.1, 0.2 and 0.3 than the exponents in the spectra of Fe + Ni. Cowsik (1973) has drawn attention to the problems of energy divergence associated

TABLE I

Cosmic ray source abundances[a] (Fe $\equiv 10^6$)

Z	Cosmic ray source			Ratio Cosmic ray source solar system
	Shapiro (1971); Webber et al. (1972)	Cartwright et al. (1973)	Shirk et al. (1973)	
He	–	$2 \cdot 10^8$	–	0.05 to 0.1
C	5.2×10^6	$5.2 \cdot 10^6$	--	0.34
N	5.7×10^5	$6.7 \cdot 10^6$	–	0.2
O	5.5×10^6	$5.7 \cdot 10^6$	–	0.2
Ne	8.7×10^5	$5.2 \cdot 10^5$	–	~ 0.5 to 1
Na	6.7×10^4	$2.6 \cdot 10^5$	–	1 to 4
Mg	1.2×10^6	$1.4 \cdot 10^6$	–	1
Al	1.1×10^5	$3.6 \cdot 10^5$	–	1 to 3
Si	8.9×10^5	$1.4 \cdot 10^6$	–	0.8 to 1.2
S	$1.8 \cdot 10^5$	$2.1 \cdot 10^5$	–	0.3
Ca	$1.3 \cdot 10^5$	$2.1 \cdot 10^6$	–	1.6 to 2.5
Fe	10^6	$1.7 \cdot 10^6$	10^6	1 to 1.7
Co + Ni	$4 \cdot 10^4$		3.8×10^4	0.7
Cu + Zn	–		$(5 \pm 1) \times 10^3$	~ 2
Ga to Mo	–		25	~ 0.5
Hg to Bi	–		2.3 ± 1.0	1 to 2
Th + U	–		2.1 ± 1.3	1 to 4 (if Grevesse, 1969) 12 to 48 (if Cameron, 1968)
$Z \geqslant 96$	–		$\leqslant 0.8$	?

[a] Rare elements such as Li, Be and B are omitted because only upper limits consistent with solar abundance can be set.

with flatter spectra of the Fe + Ni nuclei and has developed a model in which the compositional changes are associated with energy-dependent leakage from source regions into the interstellar medium.

Table I gives source abundances calculated for nuclei that are not masked by secondaries resulting from spallation of their neighbors. Different workers now agree reasonably well on the source abundances of the more abundant elements. There is still some question whether the source abundances of Na, Al and P relative to Si are consistent with solar system abundances (Webber *et al.*, 1972) or whether they are more abundant by a factor of about three (Cartwright *et al.*, 1973). For those cosmochemists who like to visualize comparative abundances on a log-log plot, I have compared cosmic ray abundances with solar abundances in Figure 7 and with abundances of energetic solar particles in Figure 8. In order to make the comparison of cosmic ray source abundances with solar system abundances most meaningful, we must now examine recent data on solar flare particles.

6. Abundance of Energetic Nuclei Emitted in Solar Flares

Until recently, all data on the composition of solar flare particles had been obtained with nuclear emulsions exposed on sounding rockets for four-minute intervals. The more abundant even-Z nuclei with energies above about 20 MeV nuc^{-1} can be

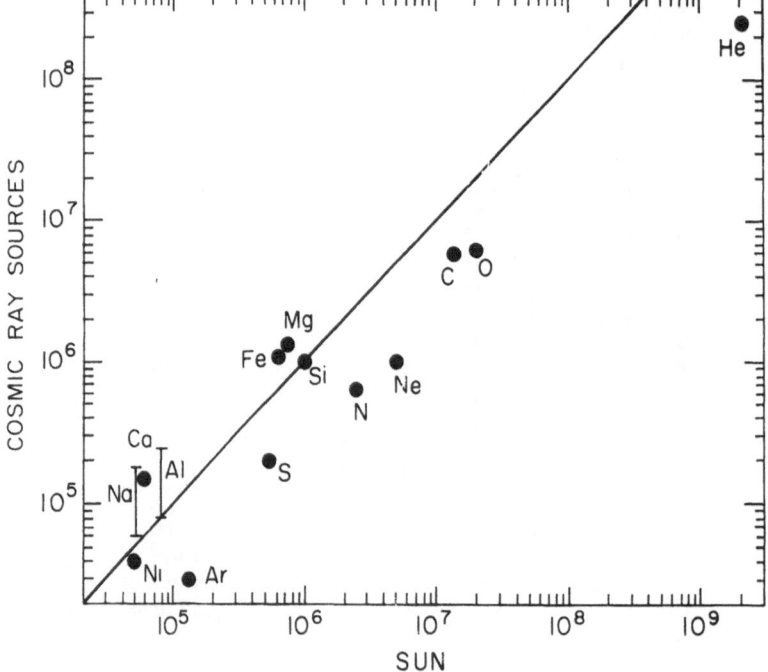

Fig. 7. Relative composition of cosmic ray sources and the Sun, normalized to Si $\equiv 10^6$. The data are from Table I.

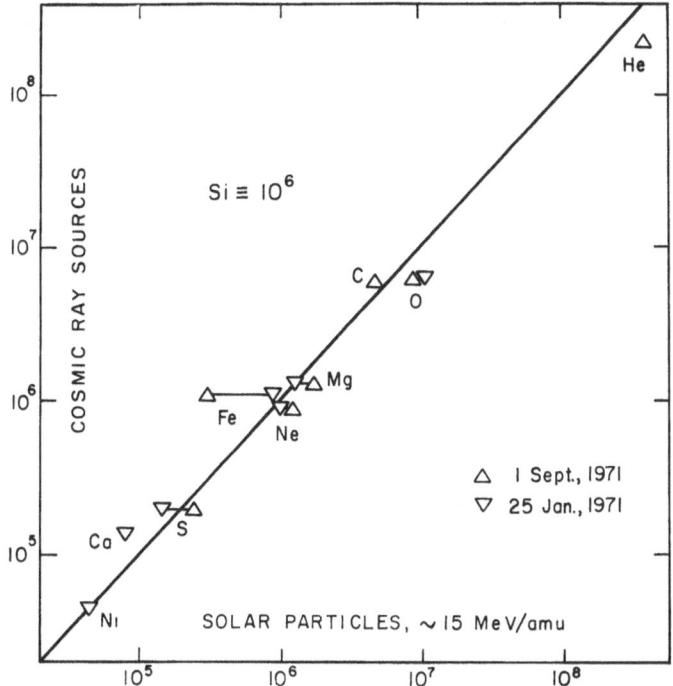

Fig. 8. Relative composition of cosmic ray sources and of energetic solar particles measured in two flares, normalized to Si ≡ 10⁶. The symbol ▽ refers to data for the 25 January 1971 event (Crawford *et al.*, 1972); the symbol △ refers to data for the 1 September 1971 event (Teegarden *et al.*, 1972).

identified. Over the years Fichtel and co-workers (Bertsch *et al.*, 1972, 1973, and references therein) have built up a consistent picture in which the relative abundances of the abundant nuclei from He to Fe appeared to be constant from flare to flare and to reflect the composition of the solar atmosphere where they originated. Their data are summarized in Figure 9. This picture has had considerable appeal to cosmochemists and astronomers because it suggests that the abundances of elements such as He, Ne and Ar that are not ionized in the photosphere can be determined from solar particle data. It has also been reassuring to cosmic ray astrophysicists because it provided evidence that nuclei with the same A/Z ratio would be accelerated without distortion of the source abundances.

Now that the capabilities exist for determining the composition of solar particles down to energies less than 1 MeV nuc^{-1} and over long periods of time, we know that the composition can vary both with time and with energy. When the Apollo 12 astronauts returned to Earth with the camera from the Surveyor 3 spacecraft that had sat on the Moon for 2.6 yr, three groups studied Fe tracks produced in the glass lens filter by solar flare particles (Crozaz and Walker, 1971; Fleischer *et al.*, 1971; Price *et al.*, 1971). Price *et al.* (1971) compared the Fe flux with the He flux measured on satellites during the same time interval and concluded that the average Fe

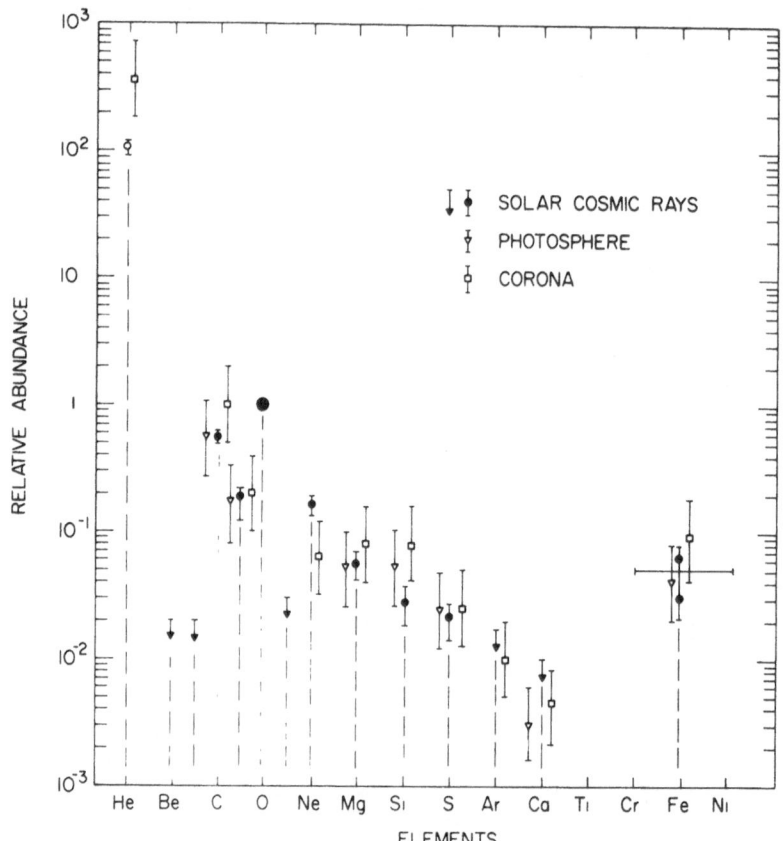

Fig. 9. Summary of nuclear emulsion data (Bertsch *et al.*, 1972, 1973) on abundances of energetic solar particles relative to photospheric and coronal values (Withbroe, 1971).

abundance relative to the He in solar flares was enhanced by a factor of ~ 10 to 20 at energies of a few MeV nuc^{-1} but approached the photospheric ratio at energies of a few tens of MeV nuc^{-1}. They suggested that heavy nuclei might be preferentially emitted from accelerating regions because of their low fractional ionization and high magnetic rigidity.

At very low energies, < 1 MeV nuc^{-1}, Armstrong and Krimigis (1971) and Van Allen *et al.* (1973), using detectors on satellites, have reported frequent fluctuations in the He to CNO ratio which appear not to occur at higher energies (Teegarden *et al.*, 1972; Armstrong *et al.*, 1972).

Crawford *et al.* (1972), using plastic detectors on board the same rockets containing nuclear emulsions belonging to Bertsch *et al.* (1972, 1973), determined the solar particle composition in the 25 January 1971 flare from O to Ni and from ~ 40 MeV nuc^{-1} down to the lowest energy yet studied, ~ 0.2 MeV nuc^{-1}. In a careful study of the two elements Fe and Si, they found that the ratio Fe/Si decreased from ~ 2 at 2 MeV nuc^{-1} to ~ 0.4 at 30 MeV nuc^{-1}. Their values of the abundances at the highest energies studied, where enhancements due to partial ionization should be absent, are compared with cosmic ray source abundances in Figure 8.

It is rather remarkable that the cosmic ray source abundances bear a much stronger similarity to solar particle abundances in that flare than to abundances in the solar photosphere. If these were the only data, one would be tempted to conclude that similar selection mechanisms operate both at the sun and at cosmic ray sources, and that nature might be selectively accelerating cosmic rays from a galactic reservoir

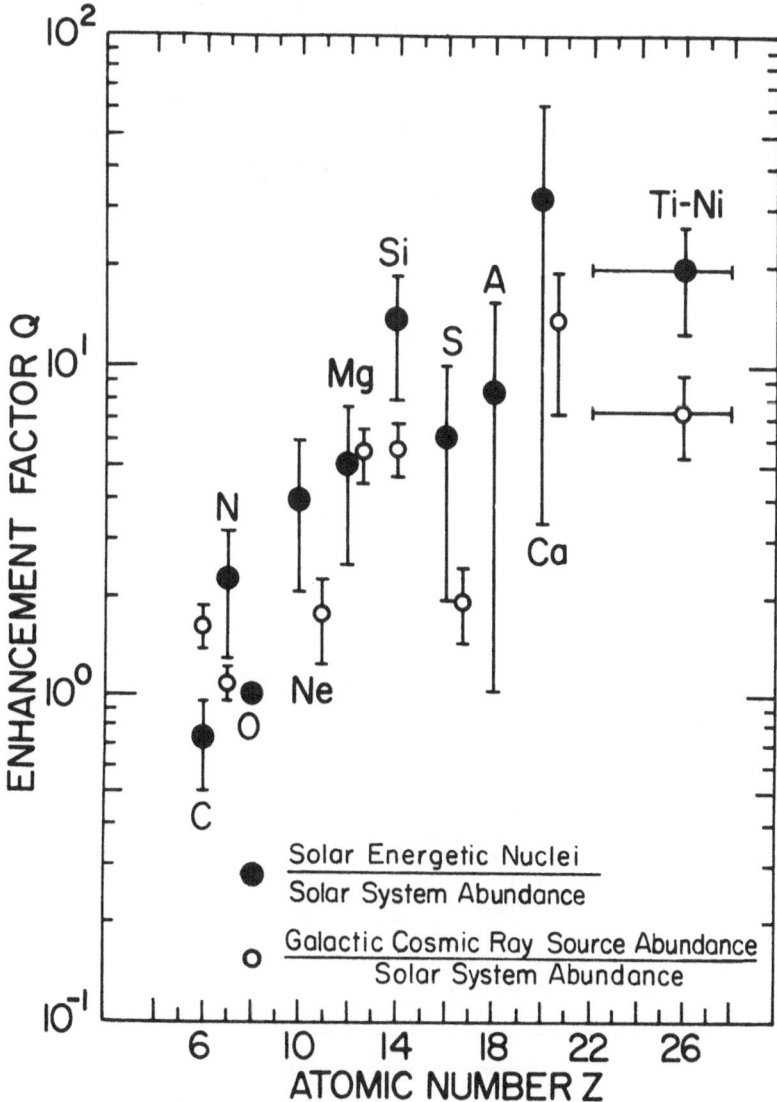

Fig. 10. Enhancement factor as a function of Z determined for solar particles averaged over seven flares (Mogro-Campero and Simpson, 1972) and for galactic cosmic ray source abundances (Cartwright *et al.*, 1973). All abundances are measured relative to oxygen.

with a solar composition in the same way that solar particles are selectively accelerated from the solar atmosphere.

We must remember that rocket data average only over a four-minute interval
during a flare, whereas cosmic rays probably represent a long-term average over
many sources. For the last year or so satellites have been providing nearly continuous
coverage of interplanetary particle fluxes. Using electronic detectors with excellent
resolution, Teegarden *et al.* (1972) have reported abundances for the entire duration
of a large flare on 1 September 1971 that are consistent with those obtained with
plastic detectors during the flare on 25 January 1971. Their data are shown in
Figure 8. Again we see a strong resemblance between abundances of solar particles
and abundances in cosmic ray sources.

In a survey of seven flares with a satellite-borne electronic detector, Mogro-
Campero and Simpson (1972) found considerable variability but generally reported
significantly higher abundances of the heavy elements $(Z \geqslant 14)$ than have been

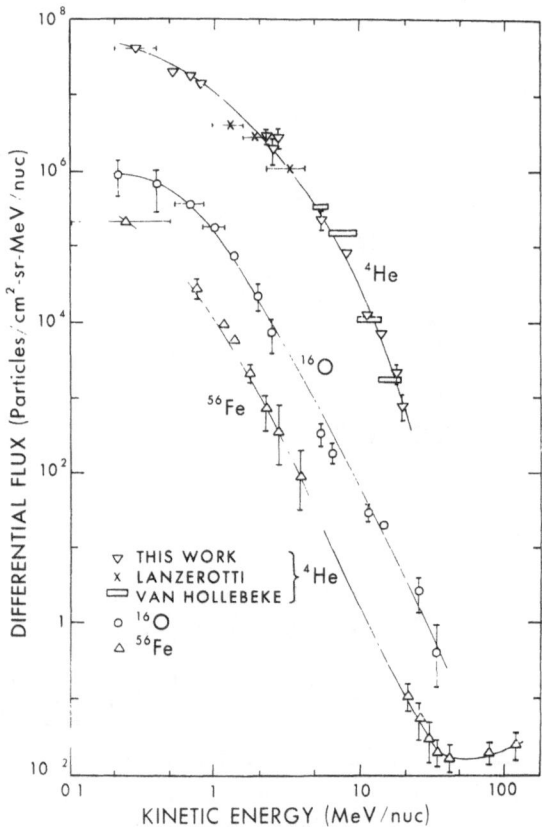

Fig. 11. Differential energy spectra of He, O and Fe summed over the flare that occurred on
18 April 1972 during the Apollo 16 mission. The data were obtained with plastic and glass detectors
(Braddy *et al.*, 1973).

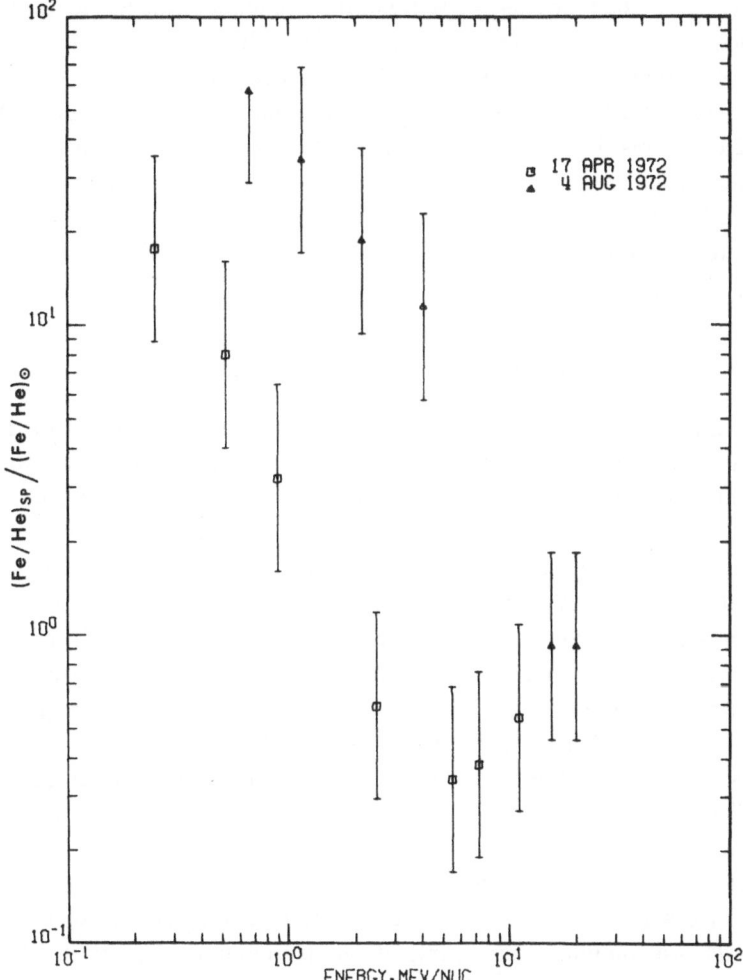

Fig. 12. Energy-dependence of the abundance ratio of He/Fe in solar flare particles relative to Fe/He in the Sun. In the intense flare of 4 August 1972 the Fe/He ratio is enhanced out to higher energies than in the weak flare on 18 April 1972. Data are from Braddy *et al.* (1973) and Price (1973).

observed with the electronic detector of Teegarden *et al.* (1972), with plastic detectors (Crawford *et al.*, 1972), and with nuclear emulsions (Bertsch *et al.*, 1972, 1973). When they summed their observations for all flares they found an enhancement relative to solar abundance that appeared to increase with Z. Their data are shown in Figure 10. At the time of writing, the reason for the large difference between their heavy element abundances and those of other workers is not understood.

By analyzing tracks produced during a week-long exposure of their plastic detector stack on the outside of Apollo 16 spacecraft, Braddy *et al.* (1973) and Price *et al.*

(1973) have determined the energy spectra of the abundant elements He, O, Ne, Si and Fe summed over an entire flare of medium intensity that lasted for two days starting 18 April 1972. Figure 11 shows some of their spectra, which decrease monotonically by about seven orders of magnitude in the energy interval $\sim$0.1 to $\sim$20 MeV nuc^{-1}. Figure 12 compares the energy-dependence of the Fe/He ratio for this flare and for the spectacular flare on 4 August 1972. Data for the latter event were obtained from tracks in a Lexan stack exposed on a rocket fired from Ft. Churchill (Price, 1973). In both flares the Fe was strongly enhanced at low energies but the energy-dependence was quite different in the two cases.

7. What Are the Sources of Cosmic Rays?

We have just seen that the ratio of heavy to light elements varies strongly at low energies and from flare to flare. From the cosmochemical viewpoint, it is worthwhile to focus attention on the composition at the highest energies for which there are good statistics, in the hope that it will bear a simple, direct relationship to the composition of the Sun. Figures 7 and 8 show that the composition of solar particles of energies above $\sim$10 MeV nuc^{-1} resembles the composition of cosmic ray sources more than it does the composition of the solar atmosphere. Though measurements at such energies are required for many more flares before a firm picture will emerge, it is interesting to note in Figure 7 that the elements depleted with respect to Fe in cosmic ray sources and also in solar particles tend to be those with high first ionization potentials. This relationship for cosmic rays has been pointed out before by Havnes (1971). The degree of depletion or enhancement is not a simple, increasing function of Z.

I would argue that it is perhaps premature to worry unduly about the few elements in Table I for which the ratio of cosmic ray source abundance to solar abundance, shown in the last column, is significantly different from unity. The main offenders are H and He (depleted by about an order of magnitude in cosmic ray sources), CNO (depleted by a factor 3 to 5), and possibly U Th (enhanced if the meteoritic abundances are correct; but normal if the solar spectroscopic values are correct). We can dismiss most of the remaining discrepancies by pointing out that uncertainties of a factor two are the norm in solar spectroscopy. Perhaps we should first tackle the easier task – that of understanding selection mechanisms by which the relative abundances of energetic particles are established in discrete solar flares – before going on to the more difficult task of studying selection mechanisms operating at cosmic ray sources.

Representing one extreme point of view is the hypothesis that the raw material of cosmic rays is gas whose composition is similar to that of our solar system and may in fact represent a good sample of the composition of our Galaxy. In the vicinity of powerful, magnetized objects such as rotating neutron stars (Gunn and Ostriker, 1969; Goldreich and Julian, 1969) or rotating white dwarfs (Cowsik and Price, 1971) the ambient interstellar gas might be ionized and swept up to relativistic energies.

The ultra-heavy cosmic rays could in principle provide a critical test of this perhaps oversimplified viewpoint. With improved resolution and collecting power it should be possible to establish limits on the abundance of long-lived trans-uranic cosmic rays sufficiently stringent to determine whether they have reached secular equilibrium between radio-active decay and continuing production in r-process synthesis. If so, the proportion of ^{244}Pu would be $\sim 2\%$ that of U, and the abundances of all other trans-uranic cosmic rays would have decayed to negligible levels. If the ratio $[Z \geqslant 94]/U$ is considerably greater than 2% then we must look to cosmic ray sources in which trans-uranic nuclei are accelerated shortly after being synthesized. We saw earlier that three out of eight cosmic rays observed with $Z > 83$ may be heavier than uranium. However, the difficulties of reliably determining both velocity and charge of these exceedingly rare, ultra-heavy cosmic rays are such that we cannot rule out the possibility that none of them is trans-uranic.

A resolution of the present large discrepancy between the solar (Grevesse, 1969) and meteoritic (Cameron, 1968) abundance of Th would be extremely valuable. In conversations with an astronomer (B. Pagel) and a meteoriticist (J. Lovering) at this Symposium I have learned of difficulties inherent in both techniques, but have not been able to decide where to place the blame for the discrepancy. The geochemists would of course be most upset if the meteoritic rather than solar value were to be seriously in error, but theoretical calculations (Seeger et al., 1965) of nucleosynthesis of the heavy elements agree with the solar spectroscopic value and are irreconcilable with the present meteoritic value. From Table I it is clear that a resolution of the Th and U abundance in favor of the meteoritic value used by Cameron would pose great difficulties for the point of view that cosmic rays are closely related to galactic abundances. One would then have to argue that our solar system has an atypical concentration of Th and U that is far below the galactic average.

At the other extreme is the point of view, long espoused by a number of astrophysicists, that cosmic rays are the relativistic debris of supernova explosions. Then the depletions of the light elements H, He, C, N, and O might reflect the highly evolved nature of the debris. So far, no quantitative predictions of the composition of the relativistic fraction of supernova debris have been made, but r-process elements might well constitute an appreciable fraction of this material. It would then be perfectly natural to find Th, U and trans-uranic nuclei in higher than solar system proportions. Assuming that the cosmic rays with $Z > 60$ have been correctly identified, it has been pointed out (Price et al., 1971; Schramm, 1972) that the relative abundances of nuclei in the Pt peak (74 to 80), in the Pb peak (81 to 83) and in the actinide peak (90 to ~ 96) are consistent with an r-process origin and a mean leakage lifetime of several million years.

A point of view intermediate between the two extremes is that cosmic rays come from a variety of sources whose relative contributions may change with energy. Although one could use the data in Figure 5 as supporting evidence for this point of view, an energy-dependent mean pathlength seems to account for the trend in composition more naturally. When adiabatic deceleration is taken into account (particles with as little as 20 MeV nuc^{-1} of energy having had several hundred MeV nuc^{-1} before entering the solar system), our detailed knowledge of cosmic ray composition spans only a factor ~300 in energy, whereas the energy spectrum extends another 10 orders of magnitude up in energy. Undoubtedly new clues will be awaiting us as we begin to reach into this new regime.

At the upper extreme of energies soon to be accessible with large magnetic spectrometers, it has been suggested that the composition of nuclei accelerated in supernova shocks or in the near field of rotating neutron stars should shift toward lower charge as a result of nuclear interactions with the intense, high-energy electromagnetic radiation around the stars (Apparao and Rengarajan, 1970; Balasubrahmanyan *et al.*, 1971; S. Colgate, private communication). We saw in Figure 5 that at the highest energies thus far studied, ~10^5 MeV nuc^{-1}, the shift is in the other direction, with Fe being relatively more abundant than at lower energies. Magnetic spectrometers to be flown on the HEAO satellite later this decade will be able to establish whether the composition at a factor ten higher energy shows the predicted effects of passage through an intense photon field.

The extremely steep energy spectrum reported for high-Z cosmic rays (Shirk *et al.*, 1973) – if verified in further work – would undoubtedly be highly significant and might even indicate that these nuclei have come from a different source or sources than the lighter nuclei. It is interesting to note that passage of cosmic rays with a given energy spectrum through matter transforms that spectrum into one with a shallower slope. One might then speculate that many of the high-Z cosmic rays are reaching us from a nearby source without having traversed as much matter as have the low-Z cosmic rays.

8. Concluding Remarks

Better measurements of the abundances and energy spectra of high-Z cosmic rays are of crucial importance. Large volumes of plastics and emulsions that have recently been exposed on balloon flights are being analyzed by groups at Bristol, Dublin, NASA Manned Spacecraft Center, and Berkeley. A one-year exposure of plastic stacks about 1.5 m^2 in area on NASA's Skylab, starting in June 1973, should greatly increase our statistics of the ultra-heavy cosmic rays.

Though we cannot yet answer the question posed in the previous section, it has become possible within the last couple of years to begin constructing a table of approximate cosmic ray source abundances. In cosmic ray studies we are thus beginning to recognize the forest, whereas in solar particle studies we are still looking at individual trees. Within another couple of years we will probably have a much clearer idea of the energy-dependence of solar particle composition. It is my conviction that continued study of the relationships between cosmic rays and solar particles will reveal patterns of direct relevance both to acceleration process and to the nature of cosmic ray sourses.

This work was supported by AEC contract AT(04-3)-34 and by NASA grant NGR 05-003-376 and NASA contract NAS 9-12005.

References

Apparao, M. V. K. and Rengarajan, T. N.: 1970, *Astrophys. Letters* **6**, 229.

Armstrong, T. P. and Krimigis, S. M.: 1971, *J. Geophys. Res.* **76**, 4230.

Armstrong, T. P., Krimigis, S. M., Reames, D. V., and Fichtel, C. E.: 1972, *J. Geophys. Res.* **77**, 3606.

Arnold, J. R., Honda, M., and Lal, D.: 1961, *J. Geophys. Res.* **66**, 3519.

Audouze, J. and Cesarsky, C. J.: 1973, *Nature Phys. Sci.* **241**, 98.

Balasubrahmanyan, V. K., Ormes, J. F., and Ryan, M. J.: 1971, *Proc. 12th Inter. Cosmic Ray Conference*, Hobart, Tasmania **1**, 344.

Bertsch, D. L., Fichtel, C. E., and Reames, D. V.: 1972, *Astrophys. J.* **171**, 169.

Bertsch, D. L., Fichtel, C. E., Pellerin, C. J., and Reames, D. V.: 1973, *Astrophys. J.* **180**, 583.

Braddy, D., Chan, J. H., and Price, P. B.: 1973, *Phys. Rev. Letters* **30**, 669.

Cameron, A. G. W.: 1968, in L. H. Ahrens (ed.), *Origin and Distribution of the Elements*, Pergamon Press, Oxford, p. 125.

Cantelaube, Y., Maurette, M., and Pellas, P.: 1967, *Symposium on Radioactive Dating and Methods of Low-Level Counting*, I.A.E.A., Monaco, Vienna, p. 215.

Cartwright, B. G., Garcia-Munoz, M., and Simpson, J. A.: 1971, *Proc. 12th Inter. Cosmic Ray Conference*, Hobart, Tasmania, **1**, 215.

Cartwright, B. G., Garcia-Munoz, M., and Simpson, J. A.: 1973, *Astrophys. J.*, to be published.

Casse, M., Koch, L., Lund, N., Meyer, J. P., Peters, B., Soutoul, A., and Tandon, S. N.: 1971, *Proc. 12th Inter. Cosmic Ray Conference*, Hobart, Tasmania **1**, 241.

Chan, J. H. and Price, P. B.: 1973, *Proc. 13th Inter. Cosmic Ray Conference*, Denver, Colorado, paper 364.

Cowsik, R.: 1973, *Proc. 13th Inter. Cosmic Ray Conference*, Denver, Colorado, paper 359.

Cowsik, R. and Price, P. B.: 1971, *Physics Today* **24**, 30.

Crawford, H. J., Price, P. B., and Sullivan, J. D.: 1972, *Astrophys. J.* **175**, L149.

Crozaz, G. and Walker, R. M.: 1971, *Science* **171**, 1237.

Fichtel, C. E. and McDonald, F. B.: 1967, *Ann. Rev. Astron. Astrophys.* **5**, 351.

Fleischer, R. L. Hart, H. R., and Comstock, G. M.: 1971, *Science* **171**, 1240.

Goldreich, P. and Julian, W. H.: 1967, *Astrophys. J.* **157**, 869.

Grevesse, N.: 1969, *Solar Phys.* **6**, 381.

Gunn, J. E. and Ostriker, J. P.: 1969, *Phys. Rev. Letters* **22**, 728.

Havnes, O.: 1971, *Nature* **229**, 548.

Juliusson, E., Meyer, P., and Muller, D.: 1972, *Phys. Rev. Letters* **29**, 445.

Maurette, M., Thro, P., Walker, R., Walker, R., and Webbink, R.: 1968, in P. M. Millman (ed.), *Meteorite Research*, D. Reidel, Dordrecht, p. 286.

Meneguzzi, M.: 1973, *Nature Phys. Sci.* **241**, 100.

Mogro-Campero, A. and Simpson, J. A.: 1972, *Astrophys. J.* **171**, L5 and **177**, L37.

Ormes, J. F. and Balasubrahmanyan, V. K.: 1973, *Nature Phys. Sci.* **241**, 95.

Ormes, J. F., Balasubrahmanyan, V. K., and Ryan, M. J.: 1971, *Proc. 12th Inter. Cosmic Ray Conference*, Hobart, Tasmania, **1**, 178.

O'Sullivan, D., Thompson, A., and Price, P. B.: 1973, *Nature Phys. Sci.* **243**, 8.

Price, P. B.: 1971, Rapporteur Paper, *12th Inter. Cosmic Ray Conference*, Hobart, Tasmania.

Price, P. B.: 1973, in *High-Energy Phenomena on the Sun*, NASA Special Publication, p. 377.

Price, P. B., Fowler, P. H., Kidd, J. M., Kobetich, E. J., Fleischer, R. L., and Nichols, G. E.: 1971, *Phys. Rev.* **D3**, 815.

Price, P. B., Hutcheon, I., Cowsik, R., and Barber, D. J.: 1971, *Phys. Rev. Letters* **26**, 916.

Price, P. B., Rajan, R. S., and Shirk, E. K.: 1971, *Proc. Second Lunar Science Conference* **3**, 2621, MIT Press.

Price, P. B., Braddy, D., O'Sullivan, D., and Sullivan, J. D.: 1973, Preliminary Science Report for Apollo 16, NASA Special Publication, p. 315.

Ramaty, R., Reames, D. V., and Lingenfelter, R.: 1970, *Phys. Rev. Letters* **24**, 913.

Schramm, D. N.: 1972, *Astrophys. J.* **177**, 325.

Seeger, P. A., Fowler, W. A., and Clayton, D. D.: 1965, *Astrophys. J. Suppl.* **11**, 121.

Shapiro, M. M.: 1971, Rapporteur Paper, *12th Inter. Cosmic Ray Conference*, Hobart, Tasmania.

Shapiro, M. M. and Silberberg, R.: 1970, *Ann. Rev. Astron. Astrophys.* **8**, 323.

Shirk, E. K., Kobetich, E. J., Price, P. B., Eandi, R., Osborne, W. Z., and Pinsky, L.: 1973, *Phys. Rev.* **D7**, 3220.

Silberberg, R. and Tsao, C. H.: 1973, *Astrophys. J. Suppl.* **25**, 315.

Simpson, J. A.: 1971: 1971, Rapporteur Paper, *12th Inter. Cosmic Ray Conference*, Hobart, Tasmania.

Smith, L. H., Buffington, A., Smoot, G. F., Alvarez, L. W., and Wahlig, M. A.: 1973, *Astrophys. J.*, to be published.

Teegarden, B. G., von Rosenvinge, T. T., and McDonald, F. B.: 1973, *Astrophys. J.* **180**, 571.

Van Allen, J. A., Venkatarangan, P., and Venkatesan, D.: 1973, to be published.

Webber, W. R., Damle, S. V., and Kish, J.: 1972, *Astrophys. Space Sci.* **15**, 245.

Webber, W. R., Lezniak, J. A., Kish, J. C., and Damle, S. V.: 1973, *Nature Phys. Sci.* **241**, 96.

Withbroe, G. L.: 1971, in K. B. Gebbie (ed.), *The Menzel Symposium on Solar Physics, Atomic Spectra, and Gaseous Nebulae*, *NBS Spec. Publ.* **353**, p. 127.

GAS AND DUST IN COMETS

A. H. DELSEMME

Ritter Astrophysical Research Center,
The University of Toledo, Toledo, Ohio 43606, U.S.A.

1. Introduction

A comet becomes a spectacular phenomenon for a transient time only. It is a starlike object at large distances from the Sun. The cometary head typically appears at distances less than 3 astronomical units (AU), and the tail at less than 1.5 AU (although exceptions are known). This suggests that the transient phenomena appear only while a strong interaction prevails between the solar light or the solar wind, and something rather elusive that we call the cometary nucleus. The nucleus, even if it is too tiny to be easily seen, must be *the* permanent feature that is the source and the origin of these transient phenomena.

During the last fifty years, the physical study of the interaction of the nucleus with the solar light and the solar wind has progressed tremendously, while the constitution of the nucleus itself has remained largely hypothetical. It is still debatable whether the bright point of light that is often seen at the center of the coma, and that we may call the 'photometric' nucleus, can be identified with one lump piece of material that would be the true nucleus.

The purpose of this review is to discuss the origin, the nature and the fate of the material, either gaseous or dusty, that appears during the transient cometary phenomena. For this purpose, the large-scale features will be shortly described first; secondly, the small-scale features surrounding the nucleus, and finally the nucleus itself.

2. The Tail

There are two types of tail that are often seen simultaneously. They are morphologically very different. Their motions and spectra suggest that they are made of a different material. One type of tail would be made of dust; the other, of plasma.

The dust tails show the reflected solar spectrum only, slightly reddened and slightly polarized; that suggests very fine dust (Remy-Battiau, 1964; Donn *et al.*, 1967).

The plasma tails are straighter and narrower and contain more fine structure, like nodules and twists (Brandt, 1968); their spectrum shows the bands of positive molecular ions; mainly CO^+, CO_2^+ and N_2^+ (Swings and Haser, 1956); weak bands of CH^+ and OH^+ can often be observed in the head; they are also attributed to the tail as it stems from the nucleus, although being too weak to be detected any further. The free electrons that are also likely to be in the tail cannot be detected.

The gross features of the dynamics of the tails are rather well understood; in

A. G. W. Cameron (ed.), Cosmochemistry, 89–101. All Rights Reserved

particular their shape, their orientation and their motions. From the tail kinematics, the magnitude of the repulsive acceleration explaining its motion is easy to compute (Bessell, 1836; Bredichin, 1903) but its nature is less obvious.

For most of the dust tails, the radiation pressure of the sun applied to fine dust gives the right order of magnitude of the observed acceleration. The small initial velocity needed to explain the kinematics (0.3 to 0.6 km s^{-1}) corresponds to dust dragged away from the nucleus by vaporization gases (Finson and Probstein, 1968; Sekanina and Miller, 1973). The suggestion that the dust can collect some electric charge (Notni, 1964; Belton, 1965) has not been needed to explain the dust distribution in comets Arendt-Roland and Bennett.

For the plasma tails, the radiation pressure is much too small to explain the huge accelerations observed (Wurm, 1943); only a magneto-hydro-dynamic interaction with the solar wind can give the right magnitude of the observed accelerations (Biermann, 1951). The lines of force of the weak magnetic field are frozen in the solar wind plasma; when they meet the cometary plasma, they exert a violent breaking action on both plasmas, which are made to move as one fluid at their surface of contact (Alfvén, 1957). Transverse plasma instabilities may also contribute to the coupling with the solar wind (Wallis, 1967). The bow-wave of the solar wind in front of the comet is now described as less prominent (Wallis, 1973; Brosowski and Wegman, 1972) than was still believed a few years ago by the Munich group (Biermann et al., 1967).

The controversy on the place and mechanism of generation of the cometary ions, and of the CO$^+$ ion in particular, has not yet been settled. More morphological arguments are given by Wurm and Mammano (1973) for an ion source in the nuclear region, while laboratory rates of dissociative recombination exclude it (Biermann, 1970). Other mechanisms have been proposed. Wallis (1973) notes that the enhanced ionization rate of the 100 eV electrons is comparable to the charge exchange rate. The escape of ions along tail streamers is probably better understood than their formation mechanism (Wallis, 1967).

3. The Coma: Nature and Velocities of Molecular Fragments

The coma is an isotropic sphere of gas steadily expanding in all directions (Yoss, 1953; Miller, 1967). The spectra provide a ready explanation: all the molecular species identified in the quasispherical comas are neutral radicals (Swings and Haser, 1956); they are thus unaffected by electromagnetic forces and diffuse through the plasma undisturbed.

At large heliocentric distances, when the coma first appears, it shows the solar continuum presumably reflected by the nucleus and by dust particles dragged away by unobserved gases. For heliocentric distances shorter than 3 AU, superimposed on the continuum, there appears in emission the molecular bands of a few radicals: C_2, CN and CH were identified at the turn of the century; OH and NH, as well as the triatomic radicals NH_2 and C_3, were identified in the 1940's and later, mainly through the work of Swings and his collaborators. Last but not least, the existence of a huge hydrogen coma was discovered from space by the OAO, and observed by the OGO for three bright comets (Code, 1972; Bertaux and Blamont, 1970).

For the sake of completion, the metallic lines that appear at shorter heliocentric distances must be mentioned. The sodium D doublet appears near 0.9 AU, and numerous metals, presumably coming from the vaporization of meteoroid dust, show up in sungrazing comets, as Ikeya-Seki (1965 VIII) (Fe, Ni, Cr, Mn, Cu, K, Ca^+ (Preston, 1967)).

All these atoms or radicals emit light through a resonance-fluorescence excitation by the solar radiation (Wurm, 1934; Swings, 1941). There is only one definite exception. The forbidden red line of oxygen, discovered in Comet Mrkos by Swings and Greenstein (1958) cannot be excited by fluorescence. The puzzle of its excitation was probably solved by Biermann and Trefftz (1964). The photodissociation of its parent molecule must produce an oxygen atom already in its excited ^1D state, ready to emit the red line.

The coma is steadily lost in space, with a moderate diffusion velocity, probably thermal. A velocity of the order of 0.5 to 1 km s^{-1} was first deduced from the expansion rate of dust halos (Bobrovnikoff, 1931). Finson and Probstein's (1968) hydrodynamic approach confirms this order of magnitude. They also show that the gas-dust interaction takes place very close to the nucleus surface. The terminal velocity of the dust, typically 0.3 to 0.6 km s^{-1} is reached within 20 radii of the nucleus. The theoretical terminal velocity of the gas remains somewhat larger than that of the dust in most cases. It is of the order of 0.7 km s^{-1} for a pure water vapor model (Delsemme and Miller, 1971b) but may reach 1.1 km s^{-1} (Mendis et al., 1972). Greenstein's effect (differences in the spectral profiles of the same molecular band, at different places in the coma) can be understood as a differential Doppler effect; velocities from 0.6 to 0.8 km s^{-1} were deduced from it by Malaise (1970) in 4 different comets. Standing in contrast, the terminal velocity of hydrogen, as deduced from the Ly-α isophotes, is of the order of 8 km s^{-1} (Keller, 1973; Bertaux et al., 1973). In a pure H_2O model, most of the dissociation into H would originate outside the thermalized region. The fact that the H velocity is higher than that suggested by equipartition of thermal energy, is only related to the energy budget of the two photodissociations ($H_2O \rightarrow H + OH$, $OH \rightarrow O + H$); so far, it seems compatible with a pure H_2O parent molecule.

4. The Quest for the Major Component of the Gas Emitted in the Coma

Most of the neutral atoms or radicals observed in the coma – as well as the ions observed in the tail – are molecular fragments that can survive only at places where collisions are virtually inexistent. It is difficult to imagine that they could have been stored for aeons within the nucleus. Wurm (1934, 1935) suggested that they originate from the solar photodissociation of unobserved but more stable 'parent' molecules, steadily produced from the nucleus. Bobrovnikoff (1942) suggests CO_2 and NH_3,

Swings (1948) adds H_2O, CH_4, N_2 and CO existing in the form of ices. Whipple (1950) proposes his icy conglomerate model of the nucleus to explain the acceleration residue of Comet Encke that was unexplained by the gravitational forces. This so-called 'non-gravitational' acceleration would come from the asymmetrical vaporization of the nuclear snows; the lag introduced by the rotation of the nucleus would explain why this reactive impulse is not exactly opposite to the Sun. This implies a rather large mass loss per perihelion passage, even if the directional factor of the vaporizations is high (Levin, 1972).

If Whipple's (1950) explanation was right, the major components of the vaporizing gases were not detected. The major features of the spectra, like C_2 or CN in the coma or CO^+ in the tail, typically lead to production rates of 10^{28} mol s^{-1} for a medium bright comet (Wurm, 1963). But 10^{30} mol s^{-1} are needed with the largest possible directional factor to explain the non-gravitational forces of the same comet. A 'medium-bright' comet is, for our purpose, a comet of the type of Halley's or Bennett's (absolute magnitude $H_{10} = 4.5$).

However, weak features in the spectra were already suspected to be major components. The 3090 Å band of OH has a very low transition probability, and it was observed from the ground despite the beginning of atmospheric absorption due to the nearness of the ozone cutoff. Arpigny (1965) estimates its abundance in comet Bester (1949k) as sixteen times that of C_2 or CN. The forbidden line of oxygen, as already mentioned, comes from already excited oxygen atoms, produced in the ^1D state by the very process of dissociation. Then the number of observed transitions is the same as the production rate of its parent molecule at the steady state: this is again 10^{30} transitions/sec for a medium bright comet (Biermann and Trefftz, 1964).

More recently, a large production rate of unobserved gas was also substantiated by Finson and Probstein's (1968) analysis, to explain the dust distribution in the tail of Comet Arend-Roland (1957 III); this analysis has been extended with the same conclusions to Comet Bennett (1970 II) by Sekanina and Miller (1973).

Finally, the discovery of the Ly-α halo from space by the OAO and the OGO, in the three bright comets of 1970, Tago-Sato-Kosaka 1969 IX, Bennett 1970 II, and Encke 1971 I (Code *et al.* 1970, 1972, Bertaux and Blamont, 1970) also implies a production rate of the order of 10^{30} atoms s^{-1} for the same type of comet. The space observation of OH suggests that it is about 200 times more abundant than C_2 or CN (Code and Savage, 1972).

From recent evidence, the production rate of OH, H and O could therefore be of the same order of magnitude, and this is two orders of magnitude larger than the next most abundant species, observed in the spectra. This suggests therefore that the common parent molecule might be water, that should then be the major constituent of the volatile fraction of the cometary nucleus.

g sThe hypothesis that water only could control the vaporization of the icy conomerate dates back from Delsemme and Swings' (1952) argument on the possible plresence of gas hydrates (clathrates of gas) in the nucleus. The vapor pressure of more volatile ices is sharply reduced by the Van der Waals bonds linking the gases within the cavities of the clathrate lattice. Whether gas hydrates exist or not, the vapor pressure of water ice and of water ice only, can explain why the molecular emissions appear in the cometary spectra near 3 AU, and not near 20 to 40 AU.

Delsemme and Miller (1970) have also shown that gas hydrates are the limiting case of gas adsorption in water snow. If the amount of other more volatile materials is less than some 15% of the amount of water, then the same result could be obtained by mere adsorption of gases in the snow; the process is however described by the same formalism, and the important point is that water still controls vaporizations.

Recent evidence seems to confirm this hypothesis. In particular, the vaporization theory, developed by Delsemme (1965) for gas hydrates and independently by Huebner (1965) for water only, shows that a pure water snow model, or a water snow model with less than 15% of other volatile material, explains numerically the sudden appearance of the coma near 3 AU, as observed in most comets, because of the theoretical sudden drop in the vaporization rate shown by water ice and water ice only at the right distance. The non gravitational effects observed in the motion of comets can also be quantitatively explained by the dependence on distance of the vaporization of water and of water only. This was first pointed out for Comet Schwassmann-Wachmann II (Delsemme, 1972a) then verified for most short period comets (Marsden *et al.*, 1973). More specifically, the observations are consistent with the theoretical sudden drop in vaporization rate shown by water ice and water ice only, somewhere between 1 and 4 AU, depending on the albedos of the nuclear model used. A third confirmation comes from the r^{-6} dependence of brightness on heliocentric distance, observed from space by Code and Savage (1972) for the hydrogen and hydroxyl comas of Comet Tago-Sato-Kosaka 1969 IX; Delsemme's (1971) early discussion was contested by Wallis (1972); however, Delsemme's (1973) improved model has confirmed his previous statement, namely: that the two brightness laws observed for H and OH imply that the vaporization of water snow, and water snow only, controls the production of both H and OH in Comet Tago-Sato-Kosaka.

Early difficulties to explain the ejection velocities of the hydrogen atoms, as deduced by Keller's (1971) analysis of the Ly-α isophotes of Comet Bennett, have now been solved (Delsemme 1972b, Bertaux *et al.*, 1973) by showing that 8 ± 2 km s^{-1} is still compatible with a pure water model.

Other recent results also point to the presence of water. Keller (1973) has obtained a dependence on heliocentric distance of the production rate of hydrogen; his $r^{-2.3}$ law is compatible with a pure water model.

Wallis (1973) has also shown that the solar wind interaction is consistent with a pure water model. Finally, the dust distribution in the tail of Comet Bennett indicates that the drag force must have been essentially due to water vapor (Sekanina and Miller, 1973).

5. The Quest for the Other 'Parent' Molecules

If *water* ice or snow really controls the surface temperature of several cometary nuclei, then minor constituents can be released only in proportion to water, as predicted on theoretical grounds (Delsemme and Miller, 1970).

According to Wurm (1934–1935) these minor constituents of the nuclear snows should be the parent molecules of those radicals that have been identified in the spectra. These molecular fragments cannot conceivably be assembled in so many ways to reconstruct molecules.

Other observational data can also help the identification; in particular, the production of a given radical takes place by the exponential dissociation of its parent molecule with a half-life τ. As the parent molecule moves away from the nucleus with a mean velocity v, the radical production takes place in an extended source having a scale length $v\tau$. Study of photometric profiles of the inner coma, like in Malaise (1965), points to scale lengths of the order of 10^4 km or less, corresponding to a half life of the parent molecules at most one or two hours for CH's, two or three hours for C_3's, three or four hours for C_2's, and in the general range of 1 to 10 h for CN's (Malaise, 1968). From brightness outbursts, Wurm (1961) and Miller (1961) deduced even a shorter upper limit of 1 h for both C_2 and CN parents (all these values are reduced to 1 AU).

The space lifetimes for the decay by photodissociation of numerous possible parent molecules were estimated (Potter and Del Duca, 1964) from laboratory measurements of absorption cross sections and the solar ultraviolet flux. The shortest lifetime found for a possible parent of CN was 18 h and the parent was mono-cyanoacetylene ($H-C\equiv C-CN$); that for C_2 was 8 h, for acetylene ($HC\equiv CH$); to give another example, methane (CH_4) dissociates into CH with a half-life larger than 28 h.

All of the candidate molecules tried by Potter and del Duca have hence failed the test, because of all of them give lifetimes that are by and large one order of magnitude too large, or even much more for some of them. Potter and del Duca propose that, if the parent molecules exist, they are probably not species stable at room temperature. They mean that they could be more complex free radicals, highly unstable when not encaged in the icy matrix of the nucleus. However, Stief and De Carlo (1965) use the fact that the C_2 Swan bands do not lead to the ground state, to deduce that the parent molecule is likely to be acetylene or an acetylene-type molecule. The same type of reasoning, used in reverse, rules out ammonia and predicts hydrazine or a hydrazine-type molecule as a precursor of NH. Stief (1972) proposes arguments for methylacetylene or allene (two forms of C_3H_4) as the parent molecule of C_3.

Obvious hints are also given by the recent discovery of many complex molecules in interstellar space. Table I lists some of the possible parent molecules that would give the observed radicals by photodissociation and that have already been observed in interstellar space. The good correspondence to all known cometary radicals' seems to suggest that cometary and interstellar material could share a similar origin.

Some of these complex molecules might still give the answer to Potter and del Duca's quest, because their UV absorption cross section has not yet been studied in the laboratory, but no good prospect is in sight.

TABLE I

Radicals observed in comets			Possible precursors observed in interstellar space			
	H		All molecules observed			
O	OH	OH$^+$	H$_2$O	CH$_2$=O	CH$_3$—OH	
	CO$^+$		CO	CH$_2$=O	CH$_3$—OH	NH=C=O
CH		CH$^+$		CH$_2$=O	CH$_3$—OH	CH≡C—C≡N
C$_2$		C$_3$		CH≡C - CH$_3$		CH≡C—C≡N
	CN			H -- C≡N	H—N=C	CH≡C—C≡N
NH		NH$_2$	NH$_3$	NH$_2$ -- CH=O		NH=C=O
N$_2^+$		CO$_2^+$		Need other precursors or mechanisms involving collisions		

To escape from this deadlock, this reviewer has proposed (Delsemme, 1968) that the observed lifetimes could possibly be attributed, not to dissociation lifetimes of molecules, but to vaporization lifetimes of icy grains.

The existence of a halo of icy particles of another type had already been investigated by Huebner and Weigert (1966). In their approach, the icy particles are very numerous and screen the nuclear surface from the solar radiation. The large optical depth of the halo drastically limits the particle size, because the scattering of submicron particles is needed to keep the evaporation going. These authors were therefore lead to an icy halo that was vanishingly small, because very small particles vaporize quickly.

To reach vaporization lifetimes of the order of 10^4 s, Delsemme's idea was that a sizeable mass of ice could be distributed into a moderate number of moderately large icy particles. Then the halo has no optical depth and can reach a very large size.

The possible existence of an icy grain halo surrounding the cometary nucleus was confirmed by a few experiments to simulate cometary conditions in the laboratory (Delsemme and Wenger, 1970).

First a peculiar behavior of the solid hydrates of gas was observed; near 200 K, they look like a granular powder with a sharp size-distribution. Besides, grains were steadily stripped from the main body of snow, and dragged away by vaporizing gases, suggesting that it is likely to happen in comets. Then, their terminal velocity is set by the interaction with vaporizing gases (Finson and Probstein, 1968) and their rate of evaporation sets the size of a halo of icy grains that is going to build up around the nucleus.

Delsemme and Miller (1969, 1970, 1971a and b) have shown that the existence and size of an icy grain halo is consistent with the photometric shape of the continuum observed in Comet Burnham (1960 II); the fact that the halo of icy grains becomes an extended source of radicals is also consistent with the photometric profile of C$_2$ in Comet Burnham. The apparent scale length for the production of C$_2$ becomes related to the size of the halo, and therefore does not give any information on the half time of the parent molecule, except that it should be very short. The limiting case is that the parent molecules might even not exist, and the observed radicals could be liberated from both the icy matrix of the nucleus and of the icy grains.

The brightness profiles of CN and C_2 in Comet Bennett (1970 II) have been studied at different heliocentric distances. They are also consistent with a halo of vaporizing grains varying in proportion with r, as theory predicts (Delsemme and Moreau, 1971, 1973).

6. Dust Grains

The icy grains are likely to be only a special case of the solid grains dragged away by gases and eventually repelled into the dust tail. Solid grains may be pure dust grains, or dust grains with an icy mantle. When the icy mantle has not yet vaporized, the grains would typically be at 200 K which is their vaporization temperature. When the icy mantle has vaporized, grains can quickly reach their radiative equilibrium temperature, which is much higher.

Recent infrared observations of the nuclear region of several comets have detected a thermal flux which seems to come from this cloud of dust particles (Maas *et al.*, 1970; Kleinmann *et al.*, 1971; O'Dell, 1971). The particle temperatures are about 50% higher than those found for black bodies in the same radiation field. This suggests iron grains like meteoritic particles, but also any small enough grain, near 0.1 or 0.2 μ, which is surprizingly close to the image we have of the interstellar grain. Perhaps after all the cometary nuclei are made by the compaction of interstellar grains soldered together by their icy mantles, that can reestablish their individuality when dragged away by the nuclear vaporization (O'Dell, 1973). Although Myer (1972) claims that he might have detected a direct infrared measurement from the nucleus of Comet Bennett, it is very unlikely that he has detected anything else than the fine cone of dust, first dragged away sunwards, like a fountain, before being repelled towards the tail by the radiation pressure. Myer's doubts can easily be removed by comparing his resolving power with the fine structure of the dust jets shown in the Atlas of Cometary Forms (Rahe *et al.*, 1969).

The best assessment of the size distribution of the dust grains, and of their mass emission rate, probably comes from Finson and Probstein's approach (1968). When comparing their results for Comet Arendt-Roland 1957 III, and Sekanina and Miller's (1973) results for Comet Bennett 1970 II, there is an encouraging agreement that for very dusty comets, the mass emission rate of dust could reach the same order of magnitude as the mass emission rate of gas. Of course, the amount of dust freed by less dusty comets could be one or two orders of magnitude lower than that of gas; the total amount of gas freed by comets can therefore be used as an upper limit of the total amount of dust they loose.

The chemical composition of the cometary dust would still be highly hypothetical, if comets had not been associated with meteor showers for more than one century (Jacchia, 1963). The chemical composition of cometary meteoroids is given by Millman (1971). They show no great divergence from the composition of non-shower meteorites, at least for Na, Mg, Ca and Fe. However, light elements (H, C, O, N) are proportionately more important in cometary meteoroids than in (non-shower) meteorites (Millman, 1971). This, associated with the fact that type I carbonaceous chondrites are the meteorites that are the least depleted in light elements, suggests that type I carbonaceous chondrites could be typical of or closely related to the chemical composition of the cometary dust. It may be significant to note that the carbonaceous chondrites contain specialized minerals which have been produced by the action of liquid water on conventional meteoritic minerals (Dufresne and Anders, 1963); thus water is likely to exist in the cometary environment. There is also good evidence that the great majority of the meteoroids encountered by the earth are cometary fragments of a fragile structure and low density (Millmann, 1971).

The reviewer has tried to assess numerically the rate of mass loss affecting the set of short-period comets, assuming that water controls vaporization, and using Kresak's (1972) diagram of the distribution of comets in the a versus e plane (a: semi-major axis; e: eccentricity) because it allows an easy correction for the effects of selection. A production rate of 50 tons of H_2O per second is found. The largest unknown is the ratio dust/H_2O. If 20% is assumed, it yields the same 10 tons of dust per second that Whipple easily obtained in 1967 from an empirical formula. Whipple (1967) has even suggested that the loss rate of Comet Encke alone has been the major support for maintaining the quasi equilibrium of the dust in the zodiacal cloud over the past thousand years. Without being so drastic, because the rate of change with time and the dust production of Comet Encke has been very uncertain, it is however undisputable that the dust lost by comets can be of the same order of magnitude as the rate needed to replenish the interplanetary dust cloud and keep it at a steady state, against all destructive and dissipative processes. Other sources of dust are unlikely. Harwit (1964) and Banderman et al. (1970) have concluded that none of the mechanisms for capture of interstellar dust is sufficiently effective and contribute much to the interplanetary dust cloud. Whipple (1971) discusses space erosion from particle impacts against asteroids.

7. Are Comets the Only Source of Interplanetary Dust?

A new source of interplanetary dust has been proposed recently; Hemenway et al. (1972) claims that some dust would come straight from the Sun.

Although it came from entirely different considerations, this seems a natural extension of Wickramasinghe's (1967) work on the graphite or silicate grains produced in, and expelled from, the atmospheres of cool stars. For many years now, Hemenway has been collecting dust particles in the upper atmosphere. Their size is mainly from 0.1 to 0.7 μ, with a d^{-1} size distribution (d, particle diameter) which does not exclude particles up to some 10 μ, with a fluffy appearance. The claimed deficiency of these particles in low atomic weight elements is submitted as the best proof of their origin from the solar atmosphere. In the sun, a fractionation could take place to condense the most refractory elements like Hf, Ta, W and compounds like HfC and TaC, between 4000 and 5000 K which could possibly imply sunspots only. The continuity of the brightness profile of the solar corona with the zodiacal light is proposed to suggest a particle flux outward from the sun. Even if it is at variance with much previous thinking, the facts and theory brought about by Hemenway make an ensemble that cannot be ignored any more. However, the core of the argument lies in the high atomic weight elements identified in tiny micron-size grains by electron-microscope techniques. Positive identifications of elements from their X-ray lines are still somehow contested by specialists, but the opacity of the samples to high energy electrons seems at least an indication that the general picture could be correct. Of course it is too early to pass a final judgment; in particular, no formation and ejection mechanism has been quantitatively described, and no production rate has ever been predicted. If this proposed solar source of dust is confirmed, it is difficult to foresee whether it could compete with the cometary production, because the average fluxes are still highly uncertain.

8. Thermal History of the Cometary Nucleus

The possible correlation between interstellar molecules and comets on one side, between comets and carbonaceous chondrites on the other side, suggests that the thermal history of the cometary nucleus could be the missing link from interstellar molecules to carbonaceous chondrites.

Oort's ideas on the existence of a huge reservoir of comets seem rather confirmed by recent work (Chebotarev *et al.*, 1972). Oort's argument is based on the approximate coincidence of the sharp maximum of the distribution in $1/a$ (observed for the orbits of the nearly parabolic comets) with those distances where stellar perturbations would best randomize the aphelion velocities. Marsden and Sekanina (1973) believe that this distance could be somewhat reduced by taking systematically into account the non-gravitational forces to recompute the initial orbits of comets before their entering the inner solar system.

Of course, observations cannot tell whether these distant regions *also* contain comets with less eccentric orbits. However, if it is assumed that the present constant supply of comets is a lasting phenomenon, then each new comet must be brought into its present orbit by a mechanism that does not destroy the sharp maximum in the $1/a$ distribution; such a mechanism is found only in the perturbation by nearby stars, that can randomize the very small velocities near aphelion, without much changing the total energy of the comet. Nezhinskij (1972) computes that the half-life of the cometary cloud under stellar perturbations is more than one billion years, confirming Oort's early assessments. Therefore, it makes sense to believe that the cometary cloud dates back from the formation of the solar system. Oort had also proposed the hypothesis that comets and minor planets could have had a common origin in the asteroid belt. The nature and the amount of volatiles present in comets seems to exclude this hypothesis, at least in its original form. Saturn's distance seems the nearest place where water ice could have survived for moderately long durations in the absence of a large gravitational field.

If the early solar nebula is considered, the problem can be even more drastic. For instance, Cameron (1972) is uneasy with the idea of cool temperatures inside his model of the nebula; therefore he would prefer to put the origin of comets right away where they still are, that is in Oort's cloud; consequently he visualizes their formation in many nebular satellites left over outside of the main solar nebula, during its early collapse.

Another possibility has been carefully investigated by Safronov (1972). He shows that the theory by which the solar system was formed from cold planetesimals also leads to the production of Oort's cloud. While giant planets were growing by accretion, many planetesimals were accelerated to escape velocity by planetary perturbations. Most of the mass was ejected into interstellar space by Jupiter, but the cometary cloud, still gravitationally bound to the sun, was created mainly by Neptune. Whipple (1972) suggests that the vestiges of such a cometary belt, if any, could still be there beyond Neptune.

The previous considerations show that we do not know enough about the origin of the comets to settle the question of a possible thermal history of the cometary nuclei, that could explain any difference in chemical composition, still to be found between comets and interstellar grains. All what can be stated is that any possible heating must have been moderate and short-lived because of the volatiles that are still present in comets. The mere presence of water ice implies that the nucleus was never heated in a vacuum at a temperature of 160 K for more than one thousand years, 130 K for more than a million years, and 100 K for the lifetime of the solar system. This sets very critical limits for any possible heating, so drastic that one wonders whether there has been any at all.

To conclude, the chemical composition of comets remains an unsolved problem. However, the reviewer does not believe that the chemistry of the non-volatile fraction will tell us much more than we have already learned from meteorites. A comparison of the elements observed in the spectra of meteors and of comets is very significant; apart from a few gaps or a few additions, the lists look alike: they both show H, C, N, O, Na, metals of the iron group and probably Mg, K and Ca. The fact that Cu was identified in one (sun-grazing) comet, and not in meteor spectra, does not look very significant.

On the contrary, if recent circumstantial evidence seems to point out *water* as one of the major volatile constituents, practically nothing is known on the way H, C, N, O are built into other volatile molecules.

Partial support by NSF Grants GP-17712 and GP-39259 is gratefully acknowledged.

References

Alfvén, H.: 1957, *Tellus* **9**, 92.
Arpigny, Cl.: 1965, *Acad. Roy. Belgique Cl. Sci.* **35**, 5.
Bandermann, L. W. and Wolstendroft, R. D.: 1970, *Monthly Notices Roy. Astron. Soc.* **150**, 173.
Belton, M. J. S.: 1965 *Astron. J.* **70**, 451.
Berg, O. E. and Gerloff, U.: 1971, *Space Res.* **11**, 225, Akademie Verlag, Berlin.
Bertaux, J. L. and Blamont, J.: 1970, *Compt. Rend. Acad. Sci. Paris Ser. B* **270**, 1581.

Bertaux, J. L., Blamont, J., and Festou, M.: 1973, *Astron. Astrophys.* **23**, 415.

Bessel, W.: 1836, *Astron. Nachr.* **13**, 185, 345.

Biermann, L.: 1951, *Z. Astrophys.* **29**, 274.

Biermann, L.: 1970, *Sitz.-Ber. Bayer Akad. Wiss.* **2**, 11.

Biermann, L. and Trefftz, E.: 1964, *Z. Astrophys.* **59**, 1.

Biermann, L., Brosowski, B., and Schmidt, H. U.: 1967, *Solar Phys.* **1**, 254.

Bobrovnikoff, N. T.: 1931, *Publ. Lick Obs.* **17**, 404.

Bobrovnikoff, N. T.: 1942, *Phys. Rev.* **14**, 172.

Brandt, J. C.: 1968, *Rev. Astron. Astrophys.* **6**, 267.

Bredichin: 1903, quoted by R. Jaegermann in *Kometenformen*, St. Petersburg.

Brosowski, B. and Wegman, R.: 1972, preprint Max Plank Institute, Munich.

Cameron, A. G. W.: 1972, CERN Meeting on the Origin of the Solar System, Nice, France.

Code, A. D. and Savage, B. D.: 1972, *Science* **177**, 213.

Code, A. D., Houck, T. E., and Lillie, C. F.: 1970, Circ. 2201, Telegrams IAU.

Delsemme, A. H.: 1965, *Coll. Internat. Astrophys.* **13**, 77, also in *Soc. Roy Sci. Liège* **12** (1966), 77.

Delsemme, A. H.: 1968, in *Extraterrestrial Matter*, Proceedings 1968 Argonne Conference, Northern
    Illinois Univ. Press, Dekalb, Ill., p. 64.

Delsemme, A. H.: 1971, *Science* **172**, 1126.

Delsemme, A. H.: 1972a, Joint Session, IAU and CERN Colloquium, Nice, France, April 1972.

Delsemme, A. H.: 1972b, Tucson Comet Meeting, Paper No. 25, p. 174, in G. P. Kuiper and E.
    Roemer (eds.), *Publ. Lunar and Planet. Lab.*, Univ. of Arizona.

Delsemme, A. H.: 1973, *Astrophys. Letters* **14**, 163.

Delsemme, A. H. and Miller, D. C.: 1969, *Bull. Am. Astron. Soc.* **1**, 339.

Delsemme, A. H. and Miller, D. C.: 1970, *Planetary Space Sci.* **18**, 717.

Delsemme, A. H. and Miller, D. C.: 1971a, *Planetary Space Sci.* **19**, 1229.

Delsemme, A. H. and Miller, D. C.: 1971b, *Planetary Space Sci.* **19**, 1259.

Delsemme, A. H. and Miller, D. C.: 1971c, *Bull. Am. Astron. Soc.* **3**, 4.

Delsemme, A. H. and Moreau, J. L.: 1971, *Bull. Am. Astron. Soc.* **3**, 281.

Delsemme, A. H. and Moreau, J. L.: 1973, *Astrophys. Letters* **14**, in press.

Delsemme, A. H. and Swings, P.: 1952, *Ann. Astrophys.* **15**, 1.

Delsemme, A. H. and Wenger, A.: 1970, *Planetary Space Sci.* **18**, 709.

Dufresne, E. R. and Anders, E.: 1963, in B. M. Middlehurst and G. P. Kuiper (eds.), *The Solar
    System* **4**, 496, Univ. of Chicago Press.

Donn, B.: 1963, *Icarus* **2**, 396.

Donn, B., Powell, R. S., and Remy-Battiau, L.: 1967, *Nature* **213**, 379.

Everhart, E.: 1972, *Astrophys. Letters* **10**, 131.

Finson, M. L. and Probstein, R. F.: 1968, *Astrophys. J.* **154**, 327 and 353.

Fullerton, L. W. and Huebner, W. F.: 1972, AAS, Divis. Planet. Sci., Hawaii Meeting.

Gerloff, U. and Berg, O. E.: 1971, *Space Res.* **11**, 397, Akademie Verlag, Berlin.

Harwit, M.: 1964, *Ann. New York Acad. Sci.* **119**, 68.

Hemenway, C. L., Hallgren, D. S., and Schmalberger, D. C.: 1972, *Nature*, in press.

Huebner, W. F.: 1965, *Z. Astrophys.* **63**, 22.

Huebner, W. F. and Weigert, A.: 1966, *Z. Astrophys.* **64**, 185.

Jacchia, L. G.: 1963, in B. Middlehurst and G. P. Kuiper (eds.), *The Solar System* **4**, 774, Univ. of
    Chicago Press.

Keller, H. V.: 1971, *Mitt. Astron. Gesellschaft* **30**, 143.

Keller, H. V.: 1973, *Astron. Astrophys.* **23**, 269.

Keller, H. V.: 1973 (preprint).

Kleinmann, D. E., Lee, T. A., Low, F. J., and O'Dell, C. R.: 1971, *Astrophys. J.* **165**, 633.

Kresak, L.: 1972, in G. A. Chebotarev, E. I. Kayimirchak-Palonskaya, and B. G. Marsden (eds.),
    'The Motin, Evolution of Orbits, and Origin of Comets', *IAU Symp.* **45**, 503, D. Reidel Publ.
    Co., Dordrecht, Holland.

Levin, B. Yu.: 1972, in G. A. Cheboturev, E. I. Kazimirchek-Polonskaya, and B. G. Marsden (eds.),
    'The Motion, Evolution of Orbits, and Origin of Comets', *IAU Symp.* **45**, 261, D. Reidel Publ.
    Co., Dordrecht, Holland.

Liller, W.: 1961, *Astron. J.* **66**, 372.

Lyttleton, N. R. A.: 1972, *Astrophys. Space Sci.* **15**, 175.

Maas, R. W., Ney, E. P., and Woolf, N. J.: 1970, *Astrophys. J.* **160**, L101.

Malaise, D.: 1965, *Coll. Int. Astrophys. Liège* **13**, 199, also *Mem. Soc. Roy. Sci. Liège* **12** (1966), 199.

Malaise, D.: 1968, Ph.D. Thesis, Univ. Liège.

Malaise, D.: 1970, *Astron. Astrophys.* **5**, 209.

Marsden, B. G., Sekanina, Z., and Yeomans, D. K.: 1973, *Astron. J.* **78**, 211.

Marsden, B. G. and Sekanina, Z.: 1973, personal communication.

Mendis, D. A., Holzer, T. E., and Axford W. I.: 1972, *Astrophys. Space Sci.* **15**, 313.

Miller, F. D.: 1967, *Astron. J.* **72**, 487.

Millmann, P.: 1971, *Nobel Symp.* **21**, 157.

Myer, J. A.: 1972, *Astrophys. J.* **175**, L49.

Nezhinskij, E. M.: 1972, in G. A. Chebotarev, E. I. Kazimirchak-Polonskaya, and B. G. Marsden (eds.), 'The Motion, Evolution of Orbits, and Origin of Comets', *IAU Symp.* **45**, 335, D. Reidel Publ. Co., Dordrecht, Holland.

Notni, P.: 1964, *Veröff. Sternw. Babelsberg* **15**, 1.

O'Dell, C. R.: 1971, *Astrophys. J.* **164**, 511 and **166**, 675.

O'Dell, C. R.: 1973, *Icarus*, in press.

Potter, A. E. and Del Duca, B.: 1964, *Icarus* **3**, 103.

Preston, G. W.: 1967, *Astrophys. J.* **147**, 718.

Rahe, J., Donn, B., and Wurm, K.: 1969, *Atlas of Cometary Forms*, NASA, Publ. 198, Washington D.C.

Remy-Battiau, L.: 1964, *Acad. Roy. Sci. Belgique, Bull. Cl. Sci., 5e S.* **50**, 74.

Safronov, V. S.: 1972, in G. A. Chebotarev, E. I. Kazimirchak-Polonskaya, and B. G. Marsden (eds.), 'The Motion, Evolution of Orbits, and Origin of Comets', *IAU Symp.* **45**, 329, D. Reidel Publ. Co., Dordrecht, Holland.

Schatzmann, E.: 1952, *Colloq. Int. Astrophys. Liège* **13**, 313.

Sekanina, Z. and Miller, F. D.: 1973, *Science* **179**, 565.

Stief, L. J. and De Carlo, V. J.: 1965, *Nature* **205**, 889 and 1197.

Stief, L. J.: 1972, *Nature* **237**, 29.

Swings, P.: 1941, *Lick Obs. Bull.* **508**, 131.

Swings, P. and Greenstein, J. L.: 1958, *Compt. Rend Acad. Sci. Paris* **246**, 511.

Swings, P. and Haser, L.: 1956, *Atlas of Representative Cometary Spectra*, Institut d'Astrophys., Liège.

Swings, P. and Page, T.: 1948, *Astrophys. J.* **108**, 526

Wallis, M.: 1967, *Planetary Space Sci.* **15**, 1407.

Wallis, M.: 1972, *Science* **178**, 78.

Wallis, M.: 1973, *Planetary Space Sci.*, in press.

Wickramasinghe, N. C.: 1967, *Interstellar Grains, in Chapman* (ed.), *Internat. Astrophys. Ser.*, Vol. 9, London.

Whipple, F. L.: 1950, *Astrophys. J.* **111**, 375.

Whipple, F. L.: 1963, in B. Middlehurst and G. Kuiper (eds.), *The Moon, Meteorites and Comets*, Univ. of Chicago Press.

Whipple, F. L.: 1967, in J. L. Weinberg (ed.) *The Zodiacal Light and the Interplanetary Medium*, NASA SP-150, p. 409.

Whipple, F. L.: 1971, in T. Gehrels (ed.), *Physical Studies of Minor Planets* NASA SP-267, p. 389.

Whipple, F. L.: 1972, in G. A. Chehotarev, E. I. Kazimirchak-Polonskaya, and B. G. Mardten (eds.), 'The Motion, Evolution of Orbits, and Origin of Comets' *IAU Symp.*, **45** 401, D. Reidel Publ. Co., Dordrecht, Holland.

Wurm, K.: 1934, *Z. Astrophys.* **8**, 281.

Wurm, K.: 1935, *Z. Astrophys.* **9**, 62.

Wurm, K.: 1943, *Mitt. Hamburger Sternw.* **8**, 51.

Wurm, K.: 1961, *Astron. J.* **66**, 361.

Wurm, K.: 1963, in B. M. Middlehurst and G. P. Kuiper (eds.), *The Solar System* **4**, 173, Univ. of Chicago Press.

Wurm, K. and Mammano, A.: 1973 (preprint).

Yoss, K. M.: 1953, *Coll. Internat. Astrophys. Liège* **4**, 32.

CHEMISTRY OF THE SOLAR NEBULA*

JOHN W. LARIMER

Dept. of Geology and Center for Meteorite Studies,
Arizona State University, Tempe, Ariz. 85281, U.S.A.

Abstract. The composition, mineralogy and texture of chondritic meteorites suggest they are relatively unaltered relicts of the condensation and accretion processes which took place in the primitive solar nebula. Chondrites thus are thought to contain a unique record of the physico-chemical conditions which prevailed at the time and place (asteroid belt) of their origin. Elemental abundance patterns are an important clue to the events and processes. Most elements can be placed in one of four groups according to their observed fractionation behavior in chondritic material: refractory, siderophile, normally depleted and strongly depleted. This grouping can be explained in terms of four events which presumably took place during cooling, condensation and accretion in the nebula. In order of inferred occurrence these are: (1) partial loss of the initial condensates rich in refractory elements at $T > 1300\,\mathrm{K}$, (2) partial loss of metallic Fe-Ni grains, perhaps because they were magnetic, at 1000 to 700K, (3) partial remelting and outgassing of the condensate (chondrule formation) at 600 to 350K, and (4) accretion, when the $P - T$ conditions controlled the volatile content (500 to 350K). Total gas pressure at the time and place of accretion is estimated to fall between 10^{-6} and 10^{-4} atm.

1. Introduction

Urey (1952a, b; 1954) made the first comprehensive attempt to infer the physico-chemical conditions in the solar nebula from the chemical composition of meteorites and planets. Today, just as in the early 1950's, the best yardstick against which to compare predicted conditions is the direct evidence from primitive meteorites. From other papers presented here, it is evident that the bulk chemistry and mineralogy of the planets must still be inferred making any interpretation somewhat less tenable.

Our direct evidence thus comes from chondrites, the most abundant and most primitive class of meteorites. Over the past ten years or so a general consensus has been reached that most properties of chondrites were established prior to their accretion into parent bodies, probably asteroids. The composition, mineralogy and texture of chondrites all point to an origin in the cooling solar nebula. Nearly all known elements are fractionated to some extent in the meteorites but can be broken down into four groups: refractories, siderophiles and two groups of volatiles referred to as 'normal' and 'strongly depleted.' This grouping implies four processes. In probable order of occurrence these are: (1) partial loss of an early condensate ($T \geqslant 1300\,\mathrm{K}$), (2) partial loss of Fe-Ni particles (1000 to 700 K), (3) partial remelting of the condensate (chondrule formation, 600 to 350 K) and (4) accretion (500 to 350 K).

We shall begin with a very brief review of meteorite terminology and key compositional, textural and mineralogical data (Section 2). A summary of the predicted chemistry in a cooling gas of cosmic composition (Section 3) is followed by a discus-

* Contribution No. 80, Center for Meteorite Studies.

A. G. W. Cameron (ed.), Cosmochemistry, 103–119. All Rights Reserved

sion of the four processes thought to have taken place: fractionation of refractories (Section 4), metal-silicate fractionation (Section 5), chondrule formation (Section 6) and accretion (Section 7). This last section will be the most heavily stressed because much new information is available.

2. Classification and Composition of Chondrites

Chondrites are stony meteorites which contain mm-sized spherules called chondrules that appear to originate as molten droplets. Chondrules are embedded in a ground-mass or matrix which usually has a somewhat finer grain size. The dominant minerals in chondrites are olivine $[(Mg, Fe)_2SiO_4]$, pyroxene $[(Mg, Fe) SiO_3]$, troilite (FeS) and metal $(FeNi)$. In the more oxidized meteorites, some iron may be present as magnetite (Fe_3O_4) and the silicates are hydrated.

Chondrites can be chemically subdivided in five groups referred to as E, H, L, LL and C. The E-group is characterized by a highly reduced mineralogy. Nearly all of

TABLE I

Classification (van Schmus and Wood, 1967)

	Class	$Fe°/Fe$	Fe/Si	Mg/Si
	E Enstatite	0.80 ± 0.10	0.83 ± 0.32	0.79 ± 0.06
	H High Fe	0.63 ± 0.07	0.83 ± 0.08	0.96 ± 0.03
Ordinary	L Low Fe	0.33 ± 0.07	0.59 ± 0.05	0.94 ± 0.03
	LL Low low Fe	0.08 ± 0.07	0.53 ± 0.03	0.94 ± 0.03
	C Carbonaceous	Low	0.89 ± 0.08	1.05 ± 0.03

their iron is present as metal or sulfide. These may be contrasted with C-group chondrites (carbonaceous) where little or no metal can be found because all the iron is present in the oxidized form as silicates or iron oxides. The three remaining groups, frequently lumped together and referred to as ordinary chondrites, contain iron in both oxidized and reduced forms. The gradation in amount of oxidized iron is not continuous. Instead, each group is characterized by its own, unique oxidation state. In addition, the total iron contents vary somewhat and they too are quantized. This gives rise to the breakdown of the ordinary chondrites into high iron ($=H$), low iron ($=L$), and low iron-low metal ($=LL$), or low, low iron. Other chemical differences have only more recently come to be accepted as fundamental. The Mg/Si ratio, for example, is also quantized being highest in C-group, intermediate in the ordinary chondrites and lowest in E-group.

Each of the five groups may be further broken down by 'petrologic type', simply numbered 1–6, allowing each meteorite to be pigeon-holed by a letter and number; C1, L3, H5, etc. (Van Schmus and Wood, 1967). The numbers were originally

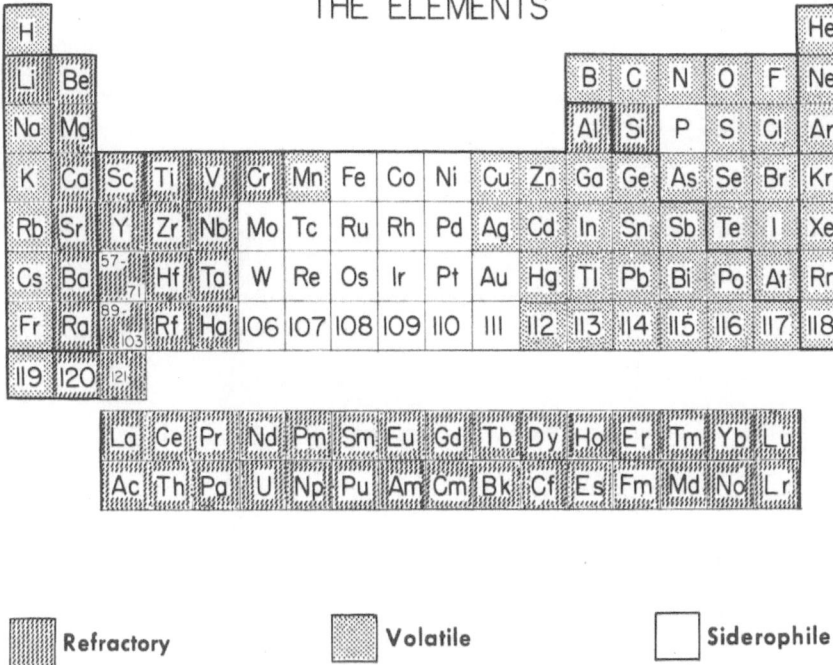

Fig. 1.   Cosmochemical classification of the elements according to their fractionation
behavior in chondrites.

assigned to reflect the extent of chondrule-matrix intergrowth (or volatile content in the case of carbonaceous chondrites). There is a growing consensus, not yet com-complete but widespread, that these textural relations are the result of metamorphic recrystallization in a meteorite parent body (Dodd, 1969). They correlate with volatile content which has led to the interpretation that volatiles were outgassed during metamorphism (Dodd, *loc. cit.*). A competing view holds that the correlation is not causal but that the volatile content was established at the time of accretion. Accordingly, the more volatile-deficient, highly metamorphosed meteorites are thought to have accreted earlier at somewhat higher temperatures and subsequently resided deeper in the parent body (Larimer and Anders, 1967).

The observed chemical fractionations in chondrites are best discussed by reference to a periodic table (Figure 1). The elements are divided here into three groups: refractories, siderophiles and volatiles. Refractories include most elements on the left side of the table plus Al, Si, the lanthanides and actinides. Most of the transition metals 'move' with iron; such elements are referred to as siderophile. The volatile elements include the alkalis and Mn plus most elements on the right side. For reasons which will become apparent. the volatiles are divided into a 'normal group', (Cu, F, Ga, Ge, S. Sb, Se and Sn) and a 'strongly depleted group', which encompasess the remainder (Anders, 1964).

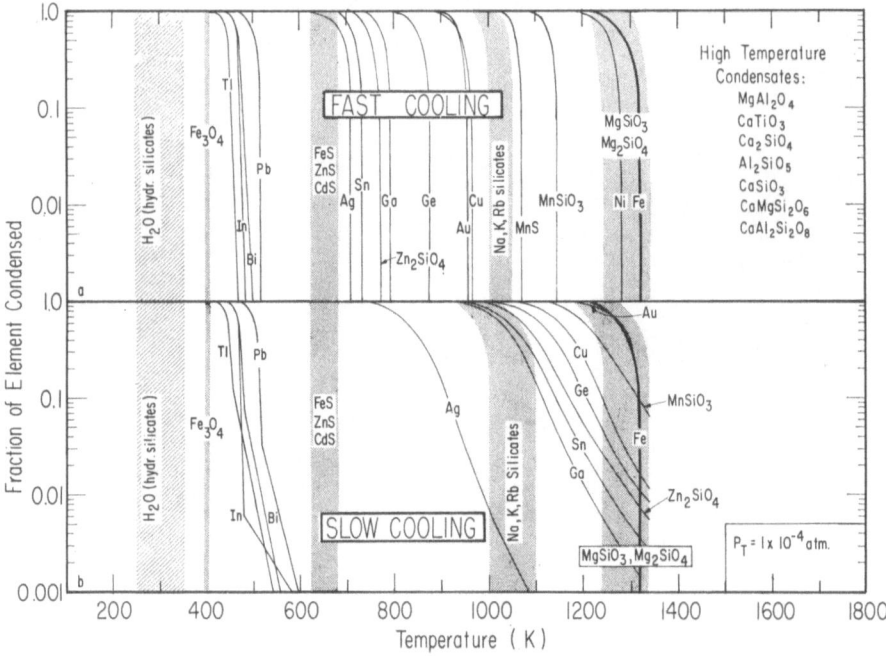

Fig. 2. Condensation sequence in a gas of cosmic composition (Larimer, 1967, with minor revisions). Lower sequence assumes equilibration, upper sequence assumes diffusional equilibrium breaks down owing to rapid cooling rate, large grain size, or both.

3. Condensation

This grouping according to fractionation behavior closely parallels the predicted condensation sequence of the elements from a cooling gas of cosmic composition (Figure 2). The first materials to condense, as first noted by Lord (1965), are a group of rare minerals characterized by high Ca, Al, Mg and Ti contents. The first major condensate is either iron or Mg-silicate (olivine, pyroxene); iron condenses first at $P_t \gtrsim 10^{-5}$ atm and the Mg-silicates condense first at lower pressures (Grossman, 1972). Once iron grains are formed, they provide a site for the condensation (as alloys) of other siderophile elements. (Note that Ga in Figure 2b is not fully condensed until less than 1000 K). At 680 K, the iron grains begin to react with H_2S (gas) to form FeS. Some key volatile trace elements, Pb, Bi, In and Tl, condense next followed by the conversion of any remaining iron grains to magnetite at 400 K ($\frac{3}{2} Fe + H_2O \rightarrow \rightarrow \frac{1}{4} Fe_3O_4 + H_2$). Near 350 K, the silicates react with H_2O in the vapor to form hydrated compounds.

For reasons which will become apparent, it is worth noting a few generalizations here. Carbonaceous chondrites contain hydrated silicates, magnetite and their full complements of the volatile trace elements implying that they accreted at or below about 350 K, as first noted by Urey in the early 1950's. Ordinary chondrites, on the other hand, contain FeS, no indusputable primary Fe_3O_4 and variable amounts of Pb, Bi, In and T1 suggesting an accretion temperature of 500 ± 100 K.

4. High-Temperature Fractionation

The first clue that a high-temperature fractionation process may have taken place is the difference in Mg/Si ratios among chondrites (Table I). It is apparent that if such a process is hypothesized, as opposed to liquid-crystal differentiation, similar fractionation of other refractories should be observed. The first test of the hypothesis was to check the Ca/Si and Al/Si ratios in chondrites (Figure 3). Again, it was found that

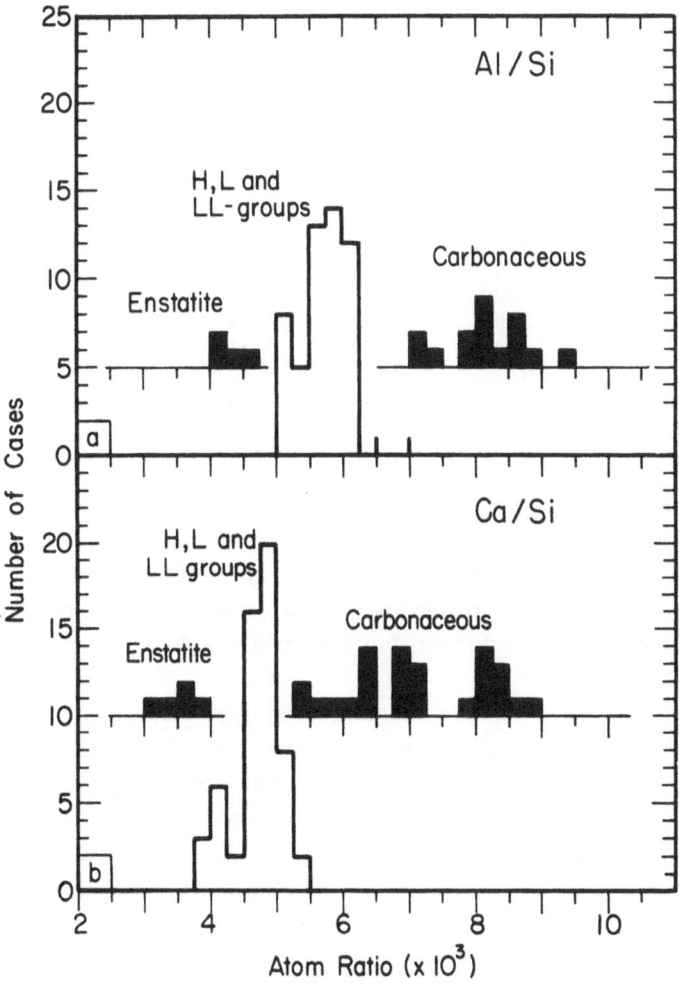

Fig. 3. Ca/Si and Al/Si ratios in chondrites decrease in the order $C > O > E$. On the basis of X-ray fluorescence measurements, Ahrens *et al.* (1969) have shown that each group has an even narrower range of ratios than illustrated here. Data from the compilation of Larimer and Anders (1970).

the ratios increase in the order $E < 0 < C$. This was followed up by a complete survey of the data on the remaining twenty-odd refractories and, while the data are not yet complete, analogous fractionations have been found in the majority of cases (Table II).

This suggests that the trend from C to E chondrites can be explained by progressively greater loss of an early condensate enriched in these elements. At about the time the data were being compiled on refractories, Christrophe (1968), shortly followed by Keil *et al.* (1969), reported the presence of cm-sized inclusions comprised of $MgAl_2O_4$, $CaTiO_3$, etc., in some carbonaceous chondrites. Also at about this time (1969, February 8) a large carbonaceous chondrite (Allende) which fell in Northern Mexico

TABLE II

Abundances of refractory elements relative to carbonaceous chondrites (Larimer and Anders, 1970)

Element	Ordinary chondrites	Enstatite chondrites
Al	0.71	0.55
Ca	0.67	0.53
Cr	0.85	0.77
Mg	0.90	0.74
Ti	0.74	0.55
Hf	0.50	0.43
La	0.80	0.43
Sc	0.72	0.63
Th	0.65	0.49
U	0.68	0.50
Y	0.75	0.44
Yb	0.81	0.48
Zr	0.58	0.47

was found to contain numerous inclusions of similar mineralogy. Somewhat fortuitously, we were thus provided with direct evidence that minerals of the right composition had not only formed but had managed to segregate themselves into cm-sized objects.

Grossman (1973) has analyzed a number of these inclusions from Allende and reports enrichments in all refractories by factors of 18–26. In addition, he has found them to be enriched in Ir, a normally siderophile element, but a highly refractory one which reinforces the argument that these objects formed at high temperatures prior to the condensation of iron. This observation also may explain the fractionation of Ir in chondrites which does not correlate as it should with other siderophile elements (Müller *et al.*, 1971).

There is circumstantial evidence that this fractionation of refractories was widespread in the solar system and may, in fact, be a universal feature common to other nebulae. Gast (1973) has argued that the upper regions of the Earth and Moon may be enriched by a factor of five or more in refractories such as Ca, Al, Ti, Sr, U,

REE, etc. Anderson (1972) now suggests that the moon may be comprised almost totally of high-temperature condensates. Moreover, Herbig (1970) has pointed out that the interstellar gas is depleted in Ca, Al and Ti but not in more volatile elements such as Na.

5. Metal-Silicate Fractionation

The variation of Fe/Si and Ni/Si ratios in primitive chondrites is direct evidence of a metal-silicate fractionation. The probable mechanism is preferential loss of metal grains in the nebula. Density differences among the inner planets provide indirect evidence of metal/silicate fractionation though this observation is open to other interpretations (Anderson, 1973; Lewis, 1973).

Larimer and Anders (1970) have attempted to estimate the temperature at the time metal became separated from silicate. They reasoned that in order for an element to be depleted along with Fe and Ni it must have condensed onto the metal grains prior to the fractionation event. Upper limits of $T \leqslant 1050°$ and $\leqslant 985 \pm 50$ K were deduced for O and E chondrites because elements less volatile than Ge correlated with iron and nickel while more volatile elements (Ga, S) did not. This upper temperature limit was based on an assumed pressure of 10^{-4} atm. There are now reasons for believing the pressure to be perhaps an order of magnitude lower for O chondrites (Section 7) dropping these upper limits by about 50 K. (Table III).

A lower temperature limit of about 680 K is derived from the Fe^{+2} content of the silicates and the absence of FeS at the time of fractionation, as implied by the regressions of Ni/Mg and Fe/Mg ratios in chondrites.

Interestingly, the upper limits lie close (even closer at the lower pressure) to the ferromagnetic Curie point of the metal in O and E chondrites. If this were just a coincidence then it is difficult to explain why fractionation processes which began 400 higher stopped and started again at just this point. If the relationship is causal, which seems quite likely, then it implies the presence of a magnetic field.

TABLE III

Fractionation of siderophile elements in ordinary chondrites

Element	L-Group/H-Group	Condensation temperature[a] (K)
Ir	0.57	1425
Fe	0.74	1210
Ni	0.68	1210
Co	0.66	1205
Au	0.61	1130
Pd	0.55	1105
Ge	0.78	1000
Ferromagnetic Curie Point		985–900
Ga	0.93 ÷ 0.15	885
S	1.07	680

[a] For 90 % Condensation at $P_t = 10^{-5}$ atm.

6. Chondrule Formation

Anders (1964) first pointed out, on the basis of good but meager data, that a large group of volatiles appeared to decrease uniformly from C1 to C2 to C3 chondrites in approximate ratios of 1:0.7:0.3. The only satisfactory explanation he could devise was that these meteorites were a blend of two components: a high-temperature material essentially devoid of volatiles and a low-temperature material which had retained its volatiles. These two components were tentatively identified as chondrules and matrix. This identification is now supported by various lines of evidence: chondrule content increases going from C1 to C3, additional data have confirmed and refined the trend, analyses indicate chondrules are depleted in volatiles and O^{18}/O^{16} variations parallel the volatile element trends (Larimer and Anders, 1967).

Ordinary chondrites display a somewhat different pattern. The eight elements in the 'normal group' are depleted relative to C1 material by constant factors of 0.25, suggesting a chondrule/matrix ratio of 75/25. But the remaining elements are depleted by much larger fractions, up to 0.001 in some cases. This depletion is interpreted as indicating the matrix in these meteorites was accreted at somewhat higher temperatures than similar material in C-group chondrites (Section 7).

The depletion of volatiles in chondrules is consistent with a high-temperature origin as indicated by the presence of glass, their sphericity and other properties. They also have a low surface area to volume ratio relative to the fine-grained matrix which would insure, and perhaps enhance, the degree of depletion. The question of how chondrules formed, (either directly from the vapor or reheating the matrix), is not crucial to the model. However, what indications exist suggest that they are derived from remelting and outgassing of matrix material, either by electric discharges (Whipple, 1966; Cameron, 1966), or a collision mechanism (Whipple, 1972; Cameron, 1973).

7. Accretion Temperatures and Pressures

In his classic papers of the early 1950's, Urey first pointed out how the volatile contents of planets and meteorites could be used as 'cosmothermometers' to infer limits on accretion temperatures. One merely assumes that the volatile content of a body is determined by the volatile content of the dust from which it accreted. The dust, of course, would contain only those volatiles which had condensed prior to the accretion event. Knowing the composition of the dust as we now see it in the form of meteorites and planets, it is possible to work backwards using classical thermodynamics to infer something about the composition, pressure and temperature of the system at the system at the time of accretion.

In the past, it has been necessary to assume a composition (solar or cosmic abundances) and pressure (usually low, $\ll 1$ atm) to calculate the temperature. We must still assume a composition; however, our confidence in cosmic abundance estimates has grown considerably in the past few years. Anders (1971) has recently argued on the basis of five lines of evidence that current estimates (Cameron, 1968) are unlikely to be in error by factors greater than 2–5 for individual elements and 1.5 and less for groups of ten or more. Errors of this magnitude lead to vanishingly small differences in the calculated P-T conditions. The total pressure on the system no longer needs to be assumed, as we shall see.

7.1. CONDENSATION

It is necessary at this point to briefly review how the thermodynamic approach works (for a more detailed discussion, see Larimer, 1967; 1972). An element begins to condense when its partial pressure, $p(E)$, equals its vapor pressure $p°$. An element's partial pressure in a cosmic gas is simply the product of its abundance relative to H_2 times the total pressure. Vapor pressure curves can be expressed as Clausius-Clapeyron relations: $\log p° = -A/T + B$ where $A \sim \Delta Hv/2.303R$ and $B \sim \Delta Sv/2.303R$. Once the element begins to condense, its partial pressure drops by a factor $1 - \alpha$, where α = fraction condensed. Thus, setting partial pressure, $(1 - \alpha)[A(E)/A(H_2)]$ P_t, equal to the vapor pressure and rearranging, we obtain:

$$\log 1 - \alpha = -A/T + B - \log P_t - \log \frac{A(E)}{A(H_2)}. \tag{1}$$

This leads to curves like those in Figure 2a, vertical lines with a slight hook on top which develops as α increases to 0.1 or so.

Since we are dealing with trace elements it is likely that they will tend to form alloys or solid-solutions with one of the condensed, major phases. Pb, Bi and T1 are predicted to behave as metals in the highly reducing gas; therefore, they should alloy with the Fe-Ni grains. This results in a different vapor pressure curve where $p' = kNp°$. N is simply the mole fraction, here taken to be $\alpha A(E)/A(Fe + Ni)$. The constant k is evaluated for solubility data according to a relation familiar to metallurgists and solution chemists: $\log N' = -\log k = \Delta H_s/2.303RT$, where ΔH_s is the heat of solution and N' is the atomic fraction at the limit of solubility. Solubility data for the trace elements in pure Fe and Ni exists, establishing limiting ΔH_s values for the Fe-Ni grains expected to be present in the nebula. Heats of solution on the order of 12–14 k-cal are determined for pure Ni and 20–23 k-cal for pure Fe. We therefore expect values of perhaps 15–18 k-cal for the real system. The resulting equation is:

$$\log \alpha/1 - \alpha = \frac{\Delta Hv - \Delta H_s}{2.303R} - B + \log P_t + \log \frac{A(Fe + Ni)}{A(H_2)}. \tag{2}$$

At the low α values of interest, the relationship reduces to a straight line which intersects the curve obtained from Equation (1) at the limit of solubility (Figure 2b). Data used in condensation equations are presented in Table IV.

The relationships for In are perfectly analogous except that In occurs as In_2S in the gas and condenses to pure InS or InS dissolved in FeS. Chemically, its condensation is described by the reaction:

$$In_2S(g) + FeS \gtrless 2InS + Fe. \tag{3}$$

The equations analogous to (1) and (2) are changed somewhat owing to the fact that

TABLE IV

Condensation equations

Pure Phase

$\log 1 - \alpha = -A'/T + b - \log P_t$

		A'	b
Pb		10000	15.10
Bi		10310	17.56
InS		12400	22.08
Tl		9590	17.12
Cds		9100	16.89

Alloy or Solid Solution

$\log \alpha/1 - \alpha = A/T - b + \log P_t$

		A	b
	H(E)	6490	9.83
Pb	L	6820	10.26
	LL	7040	10.72
	H(E)	6700	11.00
Bi	L	7030	11.52
	LL	7250	11.98
	H(E)	6200	10.64
Tl	L	6530	11.07
	LL	6750	11.53
Cds		7135	11.51

$\log \alpha^2/1 - \alpha = A/T - b + \log P_t$

InS		4000	9.34

here the pressure of $In_2S(g)$ becomes equal to $\frac{1}{2}$ the cosmic abundance in In and the activity term is now squared in keeping with the form of the reaction depicting condensation.

This squared term leads to a somewhat less steep slope for the solid-solution portion of the InS condensation curve as compared to the similar portions of the curves for the remaining elements (Figure 2b).

It thus becomes possible to predict in considerable detail the expected abundance patterns among the elements if these curves were followed during condensation. For example, if we follow the In-Tl pair down the temperature scale, we expect at high temperatures (low Tl and In contents) to see a gradual increase in the Tl/In ratio owing to the difference in slopes between their respective curves. At some point, pure InS will begin to condense resulting in a sharp increase in In content and a corresponding sharp drop in the Tl/In ratio. This sharp break will be followed by a second increase in the Tl/In ratio because In contents, after abruptly increasing, level off while the Tl content is predicted to continually increase with further temperature decline.

Each of these seemingly unique features is evident in a plot of the In-Tl data (Figure 4). Superimposed on the data are several curves illustrating the uncertainty in the heat of solution data. The curve which gives the best fit yields ΔH_s values of 15 and 19 k-cal for Tl and InS, just about what is expected.

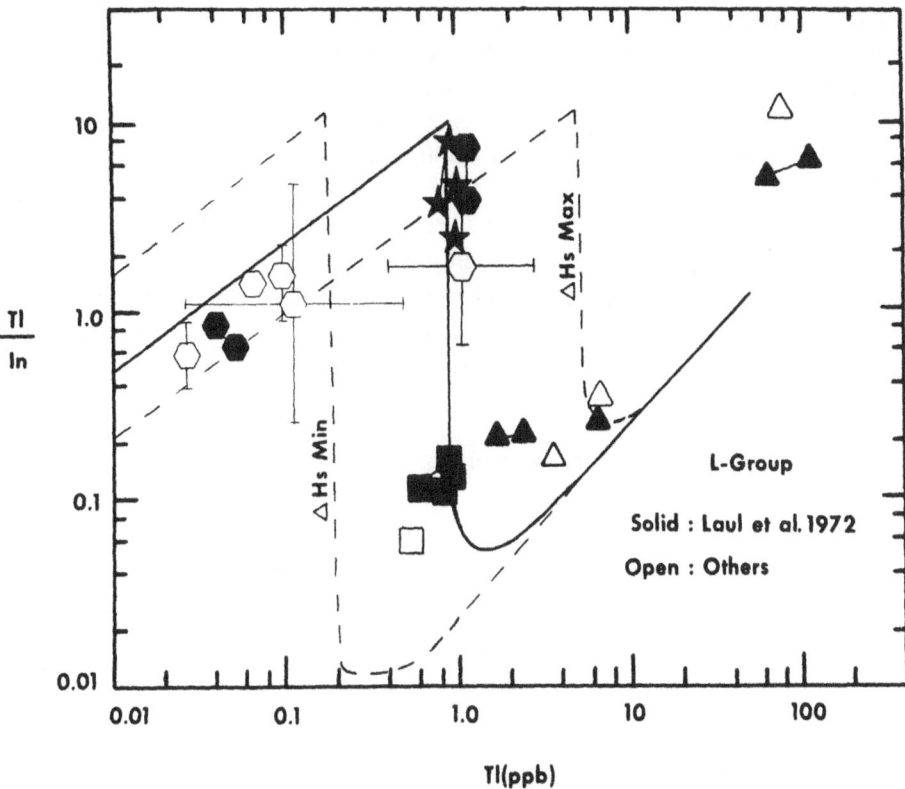

Fig. 4. Relative and absolute amounts of In and Tl in L-group chondrites tend to parallel predicted correlation curves. It seems highly improbable that non-equilibrium during condensation or redistribution during a later metamorphic event would result in identical correlations. The presently observed concentration of these elements thus appears to have been established at the time of accretion in the cooling solar nebula. Uncertainty in heat of solution ($\Delta H_s = \pm 1.5$ k-cal) is illustrated by dashed curves. Symbol morphology represents petrologic type: triangle = type 3, etc.

7.2. PRESSURE

In Equations (1) and (2) above, α is calculated from the measured trace element leaving P_t and T as the two variables. Therefore, before attempting to estimate accretion temperatures from the trace element contents it would be useful to estimate the pressure.

For ordinary chondrites, upper and lower pressure limits at the time and place of accretion can be estimated by reasoning as follows. Increases in pressure cause the condensation curves of the trace elements in Figure 2b to shift to higher temperatures, decreases in pressure cause them to shift to lower temperatures. However, their position on the temperature scale is constrained to fall between the formation temperatures of FeS and Fe_3O_4 because ordinary chondrites contain their full complement of S (as FeS) but few, if any, contain Fe_3O_4. About 80% of the sulfur will be condensed at 600 K, which we take as a safe upper limit for accretion.

Neither Pb nor Bi α-values lower than 0.01 or 0.001 have ever been observed. Substituting these α-values into the Pb and Bi equation along with the upper temperature limit, 600 K, yields an upper pressure limit of about 10^{-3} atm. On the other hand, α-values for Tl close to 1 (and sometimes exceeding 1) are frequently observed in meteorites lacking Fe_3O_4. Substituting $T \geqslant 400$ K and $\alpha \geqslant 0.9$ into the Tl condensation equation gives a lower pressure limit of 10^{-6} atm.

The element pair Bi–In is of interest here also. At $P_t \sim 2 \times 10^{-5}$ atm, their condensation curves coincide; at higher pressures Bi condenses first and at lower pressure. In condenses first. Thus, material which accreted below $P_t \sim 2 \times 10^{-5}$ atm (and $T \sim 460$ K) should have Bi/In ratios less than unity. On the other hand, material which accreted above this intersection point should never have Bi/In ratios less than 1.

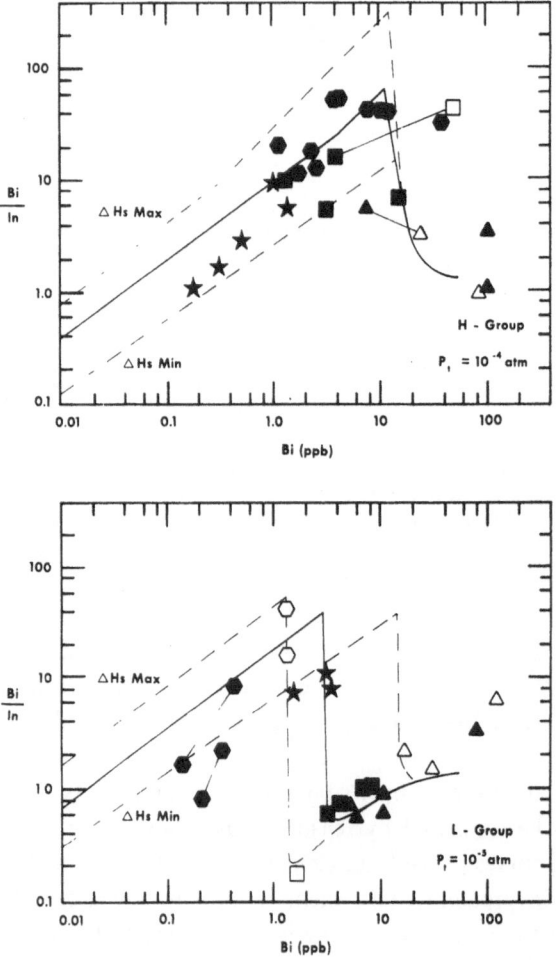

Fig. 5. Bi/In ratios frequently drop to $\leqslant 1$ in L-group chondrites but never in H-group. Evidently, pure InS condensed first in the region where L-group formed and pure Bi in the region where H-group formed. This implies a slight pressure difference, L-group at $P_t \leqslant 10^{-5}$ and H-group at $P_t \geqslant 10^{-5}$ atm.

A compilation of all reliable data from H and L-group chondrites indicates that there are numerous cases where Bi/In $\leqslant 1$ are found in L-group chondrites but none among the H-group (Figure 5). In fact, at Bi contents of 1 to 10 ppb where the ratio drops below unity for the L-group, it rises to almost 100 in the H-group, just as would be predicted if Bi condensed first (at higher P_t) in the case of the L's. It thus appears that a slight difference in P-T conditions during the accretion of these meteorites can be resolved.

7.3. ACCRETION TEMPERATURES

All available trace element data from the literature have been used to calculate the accretion temperatures shown in Figure 6. Pressures of 10^{-5} atm for L and LL-groups and 10^{-4} atm for H and E-groups were assumed on the basis of the inferences regarding a slight pressure difference.

The temperatures deduced from all three 'thermometers' indicate that a large percentage of each chondrite family accreted over a fairly narrow temperature range. Exactly how narrow is debatable; conceivably some chondrites could be a mixture of materials formed at different temperatures and what we are seeing in the figure is simply an average. However, it seems unlikely that pieces formed at grossly different temperatures ($\geqslant +10$ to $15°$ could be mixed together and still produce the type of trend shown in Figures 4 and 5 (Laul *et al.*, 1973). The 'averaged temperatures' are

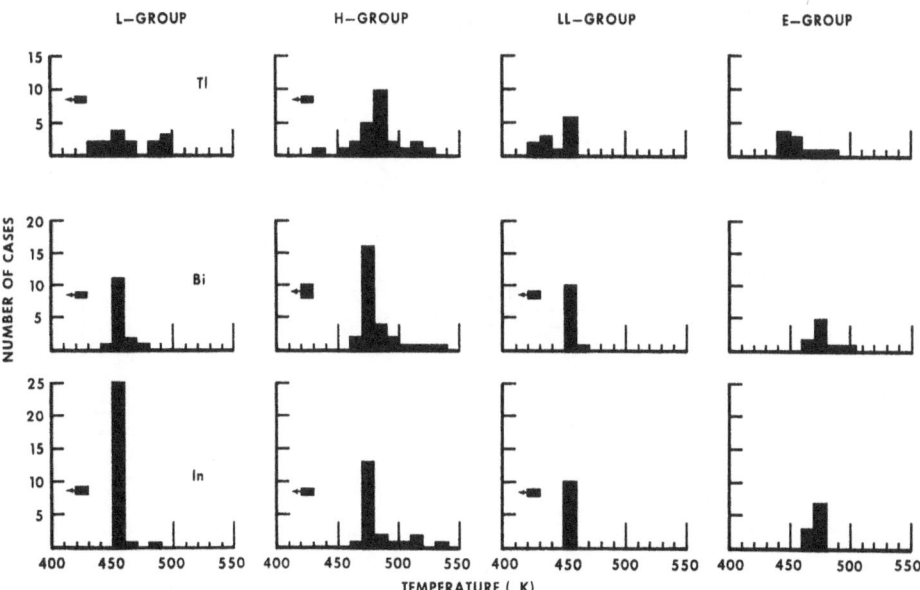

Fig. 6. Calculated accretion temperatures for the various chondrite groups tend to cluster in narrow temperature intervals. This implies that a large fraction of each chondrite parent body accreted under its own resolvable P-T conditions. The temperature for the L and LL-groups are calculated assuming $P_t = 10^{-5}$ while the H and E-groups are calculated assuming $P_t = 10^{-4}$. This gives rise to the apparent $20°$ difference in temperature.

455 ± 5 K for the L-group and 475 ± 5 K for the H-group (at $P_t = 10^{-5}$ and 10^{-4} atm).

Laul *et al.* (1973) have recently made a discovery which may have some important implications. In about 7% of the meteorites they investigated, enrichments of T1, Bi and Ag were discovered. Most such meteorites are light-dark structured, brecciated chondrites. They have proposed that the excess T1, Bi and Ag was introduced into these meteorites, probably related to the brecciation event(s), via a low-temperature, volatile-rich phase for which they coined the term 'mysterite.' For those meteorites containing this material, trace element contents cannot be considered a very reliable guide to accretion temperatures.

Accretion temperatures quite similar to those based on trace element content have been deduced from O^{18}/O^{16} thermometry (Onuma *et al.*, 1972). When the trace element and O^{18}/O^{16} temperatures are calculated for individual meteorites, they coincide almost exactly (Figure 7). In fact, the agreement almost exceeds expectations. It should be tempered somewhat, however, by the observation that most O^{18}/O^{16} temperatures for H-group condrites (in most of which trace elements have not been measured) point to a temperature of 465–470 K, whereas most trace element data

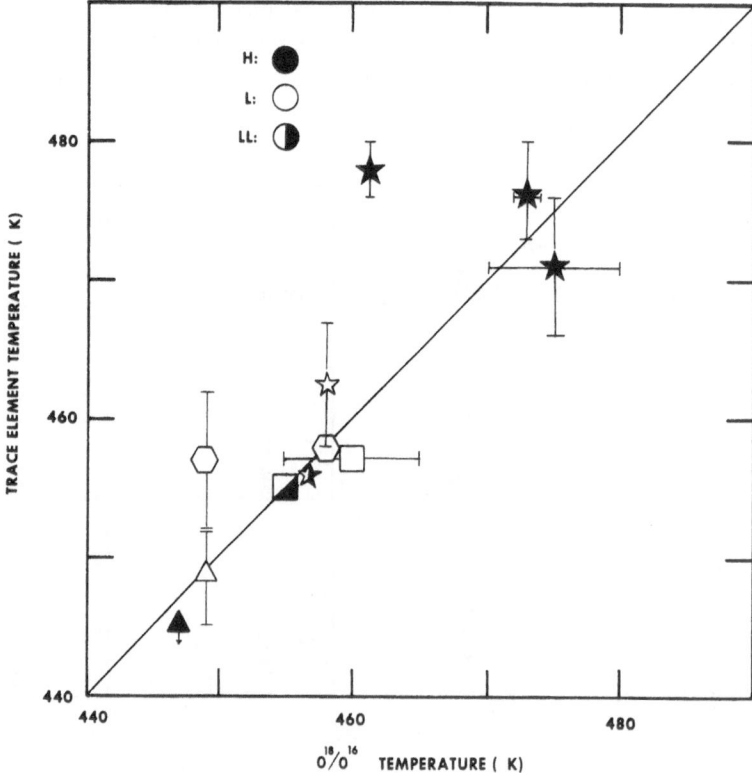

Fig. 7. O^{18}/O^{16} and trace element based accretion temperatures are highly concordant. Oxygen isotope based temperatures are pressure independent in contrast to the trace element temperatures. Concordance is achieved assuming a pressure difference by a factor of 5 to 10 between L and LL $(P_t \sim 10^{-5}$ atm) and H$(P_t \sim 5 \times 10^{-5}$ to 10^{-4} atm).

points to 470–480 K (for $P_t = 10^{-4}$). As first noted by Onuma et al., if a large percentage of each chondrite group formed over a fairly narrow temperature range it might account for their quantized oxidation states. The oxidation state of silicates in equilibrium with a gas can be obtained from the relation (Larimer, 1973):

$$\log N(\text{Fe}_2\text{SiO}_4) = \frac{1550\,(\mp 0.5)}{T} - 0.92\,(\mp 0.5) - \log\frac{\text{H}_2}{\text{H}_2\text{O}}$$

where $N(\text{Fe}_2\text{SiO}_4)$, the mole fraction of iron-olivine, is virtually equivalent to the $\text{Fe}^{+2}/\text{Fe}^{+2} + \text{Mg}$ ratio of the silicates. Substituting the observed N values of 0.19 and 0.25 for H and L chondrites into the equation, along with the predicted $\text{H}_2/\text{H}_2\text{O}$ ratio of 640 (Grossman, 1972), temperatures of 515^{+100}_{-60} and 495^{+100}_{-60} are calculated. The large error arises from uncertainty in the thermodynamic data; however, the difference in temperature $20 \pm 5°$ is subject to much smaller error and corresponds almost exactly to the temperature difference inferred from the trace element and O^{18}/O^{16} thermometers.

Additional thermometers and barometers are available for carbonaceous chondrites (C^{13}/C^{12}, solubility of Ar in Fe_3O_4, presence of other phases). These have recently

TABLE V

Formation temperatures of Cl Chondrites (Anders, 1972)

Thermometer/barometer		T(K)	P(atm)
O^{18}/O^{16}	$\text{CaMg(CO}_3)_2 - \text{H}_2\text{O}$	360 ± 5	Indep.
O^{18}/O^{16}	silicate -- H_2O	360 ± 15	Indep.
C^{13}/C^{12}	$\text{CaMg(CO}_3)_2 - $ polymer	357 ± 21	Indep.
$3/4\ \text{Fe} + \text{H}_2\text{O} \rightarrow 1/4\ \text{Fe}_3\text{O}_4 + \text{H}_2$		$\leqslant 400$	Indep.
Olivine $+ \text{H}_2\text{O} \rightarrow$ Hydr. Silicate		~ 350	10^{-4}
Tl Abundance		< 394	10^{-6}
Ar in Magnetite		350	10^{-6}

been reviewed by Anders (1972). The results are that the thermometers, both pressure dependent and independent, point to temperatures in the neighborhood of 350 to 380 K, which in turn implies a pressure of $\sim 10^{-6}$ atm. (Table V). He has also reviewed the data from the Earth, Moon and achondrites. In all of these objects, both O^{18}/O^{16} and trace element thermometry point to temperatures in the 400 to 500 K range: Moon 460 to 490 K; Earth 450 to 480 K; achondrites 430 to 480 K.

8. Conclusions

(1) Highly refractory material, rich in Ca, Al, Ti, etc., moved around in the inner solar system, presumably at $T \geqslant 1300$ K. Temperatures at least this high are thus implied for the inner part of the nebula.

(2) Metal and silicate become fractionated from one another, definitely in the case of meteorites and quite probably throughout the inner solar system. The inferred temperature limits of about 1000 to 700 K point to the possibility that the fractionation was related to the magnetic properties of the metal grains (Curie point $=900$ to 950 K).

(3) A large percentage of each group of chondrites accreted over a fairly narrow temperature range, probably on the order of ± 10 to 20 K. Most ordinary chondrites accreted in the temperature range of 450 to 480 K; the LL and L meteorites fall near the lower limit and the H near the upper limit.

(4) A variety of cosmothermometers and barometers (volatile trace elements, isotopic ratios, oxidation state, and mineralogy) yield reasonably concordant temperatures and pressures for the regions in which the bulk of each chondrite group accreted.

Group	P(atm)	T(K)
E	10^{-4}	470–500
H	10^{-5}–10^{-4}	470–480
L	10^{-5}	450–460
LL	10^{-5}–10^{-6}	440–460
C	10^{-6}	350–380

References

Ahrens, L. H., von Michaelis, H., Erlank, A. J., and Willis, J. P.: 1969, in P. M. Millman (ed.), *Meteorite Research*, D. Reidel Publishing Company, pp. 166–173.

Anders, E.: 1964, *Space Sci. Rev.* **3**, 583–714.

Anders, E.: 1971, *Geochim Cosmochim. Acta* **35**, 516–522.

Anders, E.: 1972, 'Physico-Chemical Processes in the Solar Nebula, as Inferred from Meteorites', *Proc. of Colloq. on the Origin of the Solar System*, Nice, France.

Anderson, E. L.: 1972, *Moon* **8**, 33.

Cameron, A. G. W.: 1966, *Earth Planetary Sci. Letters* **1**, 93–96.

Cameron, A. G. W.: 1968, in L. H. Ahrens (ed.), *Origin and Distribution of the Elements*, Pergamon, pp. 125–143.

Cameron, A. G. W.: 1973 'Accumulation Processes in the Primitive Solar Nebula', *Icarus*, in press.

Christophe, Michel-Levy: 1968, *Bull. Soc. Fr. Mineral. Cristallogr.* **91**, 212–214.

Dodd, R. T.: 1969, *Geochim. Cosmochim. Acta* **33**, 161–203.

Gast, P. W.: 1973, Presented at this Symposium

Grossman, L.: 1972, *Geochim. Cosmochim. Acta* **36**, 597–619.

Grossman, L.: 1973, 'Refractory Trace Elements in Ca-Al Rich Inclusions in the Allende Meteorite', *Geochim. Cosmochim. Acta*, in press.

Herbig, G. H.: 1970, *Mem (8ᵉ) Soc. Roy. Sci. Liège* **19**, 13–26.

Keil, K., Huss, G. I., and Wiik, H. B.: 1969, in P. M. Millman (ed.), *Meteorite Research*, D.Reidel Publishing Company, p. 217.

Larimer, J. W.: 1967, *Geochim. Cosmochim. Acta* **31**, 1215–1238.

Larimer, J. W.: 1973, 'Chemical Fractionations in Meteorites – VII': Cosmothermometry and Cosmobarometry', *Geochim. Cosmochim. Acta* **37**, 1603–1624.

Larimer, J. W. and Anders, E.: 1967, *Geochim. Cosmochim. Acta* **31**, 1239–1270.

Larimer, J. W. and Anders, E.: 1970, *Geochim. Cosmochim. Acta* **34**, 367–388.

Laul, J. C., Ganapathy, R., Anders, E., and Morgan, J. W.: 1972, *Geochim. Cosmochim. Acta* **37**, 329–358.

Lewis, J. S.: 1973, Presented at this symposium.

Lord, H. C., III: 1965, *Icarus* **4**, 279–288.

Müller, O., Baedecker, P. A., and Wasson, J. T.: 1971, *Geochim. Cosmochim. Acta* **35**, 1121–1137

Onuma, N., Clayton, R. N., and Mayeda, T. K.: 1972, *Geochim. Cosmochim. Acta* **36**, 169–188.

Van Schmus, W. R. and Wood, J. A.: 1967, *Geochim. Cosmochim. Acta* **31**, 747–766.

Urey, H. C.: 1954, *Astrophys. J. Supp. No. 6* **1**, 147–173.

Urey, H. C.: 1952a, *The Planets*, Yale University Press.

Urey, H. C.: 1952b, *Geochim. Cosmochim. Acta* **2**, 267–282.

Whipple, F. L.: 1966, *Science* **153**, 54–56.

Whipple, F. L.: 1972, 'On Certain Aerodynamic Processes for Asteroids and Comets', in A. Elvius (ed.), *From Plasma to Planet, Nobel Symp.* **21**, in press.

DEUTERIUM IN THE PROTOSOLAR NEBULA*

HUBERT REEVES

CSNSM, Faculté des Sciences, B.P. No. 1, 91 Orsay, France

Abstract. A number of observations, with more or less direct relationship to the abundance of deuterium, are used to evaluate the protosolar abundance of this isotope. The implications of this evaluation in the field of cosmology, stellar evolution and origin of the solar system are briefly discussed.

The problem of the determination of deuterium abundance in the primitive solar nebula can be approached by several avenues.

We may first analyze (Geiss and Reeves, 1971 hereafter GR) the meteoretic and Earth D/H water ratio $\simeq 6 \times 10^{-5}$ (Urey *et al.*, 1932; Craig, 1961; Boato, 1954). This ratio can be related to the protosolar D/H through the study of the exchange molecular reactions

$$HD + H_2O \rightleftharpoons HDO + H_2$$

if we make the reasonable assumption that during its formation, water was kept in contact with hydrogen molecules long enough for this exchange to reach equilibrium. We obtain D/H (water)$=$K $(T,$ water$) \times$D/H (hydrogen), where K $(T,$ water) is a function of temperature only, (tabulated in Reeves and Bottinga, 1972). The relevant temperature is that prevailing when water ceased to be in contact with the hydrogen reservoir.

Since water is found in carbonaceous chondrites, it may seem appropriate to choose the silicate formation temperature in these meteorites as determined from carbon and oxygen isotopic fractionation. According to Anders (1972) T should be between 360 and 400 K, for which K ranges from 2.8 to 2.2. The corresponding D/H ratio in hydrogen is $6 \pm 1 \times 10^{-5}$; which is our first estimate of the protosolar gas value.

Second we consider the observational upper limit on interstellar deuterium D/H$<7 \times 10^{-5}$, from unsuccessful attempt at detecting its 96 cm line (Weinreb, 1962). This limit would apply for protostars of today but may not apply to the protosun since we have good reasons to believe that the interstellar D fractional mass can only decrease with time; the passage of matter in and out of star (astration of matter) result in a net destruction of D. The effect of astration on interstellar D between the birth of the Sun and today has been estimated by Truran and Cameron (1972) (hereafter TC). The ratio D/H should have decreased by a little less than a factor of two according to their models. Thus we estimate D/H $\leqslant 10^{-4}$ in the protosolar nebula.

Third we consider the present solar helium isotopic ratio as obtained from the solar

---

* With an Addendum on Recent Developments in 'Deuteronomy', Dec. 1972.

A. G. W. Cameron (ed.), Cosmochemistry, 121–126. All Rights Reserved
Copyright © 1973 by D. Reidel Publishing Company, Dordrecht-Holland

wind (Geiss *et al.*, 1970, 1971) (4×10^{-4}) or the solar surface (Hall, 1972). $(4 \pm 2 \times 10^{-4})$.

The relevance of this observation stems from the fact that the present solar ^{3}He is actually the sum of two components, one of original protosolar ^{3}He, and one resulting from the fusion of the original protosolar D, through the $D + p \rightarrow {}^3$He thermonuclear reaction during the Pre-Main Sequence phase of the Sun.

We write:

$$\left.\frac{^3\text{He}}{^4\text{He}}\right|_{\text{Sun}} = \left.\frac{D}{H}\frac{H}{^4\text{He}}\right|_{\text{proto}} + \left.\frac{^3\text{He}}{^4\text{He}}\right|_{\text{proto}} = 4 \times 10^{-4}$$

and hence

$$D/H|_{\text{proto}} \leqslant 4 \times 10^{-5}$$

(where a value ^{4}He/H $|_{\text{protosolar}} = 0.1$ has been used, coherent with most observation of this ratio anywhere (Danziger, 1970)). This upper limit is lower than our estimate from the analysis of water, and suggest that our estimate of the temperature should be lowered. The same reasoning also gives an upper limit of the protosolar helium ratio $(<4 \times 10^{-4})$ slightly lower than the present interstellar upper limit $(<7 \times 10^{-4})$ (Predmore *et al.*, 1971).

Unfortunately, we cannot, this way, identify the relative contribution of the proto-solar D and ^{3}He components. An attempt at interpreting the abundances of isotopic ratios of rare gases in various rocks and meteorites has lead Black (1971, 1972) to the conclusion that he has identified rare gases implanted in meteorites before the D burning phase of the Sun. The corresponding ^{3}He/^{4}He $\simeq 1.2 \times 10^{-4}$ could then be assigned to the protosolar helium ratio, and by substraction we would have had $D/H \simeq 3 \times 10^{-5}$.

However this identification is not fully convincing; it raises several difficulties, discussed at the Nice Symposium on the 'Origin of the Solar System' (1972).

Fourth: bands of deuterated methane on the Jovian atmosphere have recently been observed by Beer *et al.* (1972). Beer and Taylor (1972) have estimated D/H to be between 3.5 and 9.8×10^{-5}.

As for water we consider the effect of molecular exchange reactions:

$$CH_3D + H_2 \rightleftharpoons CH_4 + HD$$

from which:

$$\frac{D/H \,(\text{methane}) = K \,(T, \text{methane}) \, D/H \,(\text{hydrogen})}{2}.$$

The values of K$(T, \text{methane})/2$ are given in Reeves and Bottinga (1972). As K$/2 > 1$ for all T, the upper value, $D/H = 8 \times 10^{-5}$, is also an upper limit for D/H in the Jovian atmosphere.

A determination of the effective equilibrium temperature is very difficult, since the temperature and consequently the rate of the exchange reactions increases rapidly with depth in the atmosphere and since the whole atmosphere is in rapid convective motions. Nevertheless, we may argue that its value can not be lower than the temperature of the emitting layers (~ 300 K) and hence we should have $2.9 \times 10^{-5} < D/H < 7.5 \times 10^{-5}$ in the Jovian atmosphere.

These four observations together with the arguments discussed here can be used to estimate the protosolar ratios of interest here. The following choice: $D/H = (2.5 \pm 1) \times 10^{-5}$, $He^3/He^4 = (1 \pm 0.5) \times 10^{-4}$, appears to present the least difficulties. This choice does not exclude much lower value for $^3He/^4He$; except for Black's identification we have no clear evidence of any ^3He on a cosmic scale: all observed ^3He may result from deuterium fusion or spallation reactions.

Three domains of astrophysics are in some ways interested in these estimates of protosolar ratio: the formation of the solar system, the evolution of the Galaxy and the formation of the universe.

For the solar system it appears that there has been little fractionation between the Sun-forming material and Jupiter forming material. In particular chemical separation by preferential escape of light elements could not have been important since it affects the D/H ratio more than any other isotopic ratio (but a bulk evaporation *before* the formation of both objects cannot be excluded this way).

The temperature at the moment of water formation was presumably 250–280 K.

For galactic evolution, the observation of D and ^3He in interstellar space contains information of prime importance. This was illustrated previously when the radio-astronomical upper limit of D was discussed. In the present state of the theory we believe that all deuterium atoms were born before the birth of the Galaxy and that the interstellar D/H ratio has been steadily decreasing with time. The ratio $(D/H)(t)/(D/H)(t_G)$ of these ratios at time t and at the birth of the Galaxy (t_G) can be equated to the fraction f of galactic matter which has never been involved in star formation (unastrated matter). In the calculation of TC, $f = 0.5$ at solar birth and $f = 0.3$ today. Detection of D in interstellar matter (here we predict $D/H \simeq 2 \times 10^{-5}$) and in young stars would be of great interest in this respect.

For ^3He the situation is somewhat more complex since we do not know in what way astration affects this isotope. ^3He is a normal product of hydrogen burning in stars, yielding $^3He/^4He = 10^{-4}$ to 10^{-2} according to the physical situations. These atoms would remain in the helium shell of Red Giant Stars, and could be reinjected in the interstellar medium through stellar winds. We know too little about the quantitative aspects of this problem to evaluate with any certainty whether astration results in a net depletion of ^3He (as in the case of D) or to a gradual increase (as for the elements heavier than He) as a function of time. A comparison of the protosolar $^3He/^4He$ value quoted in this paper and the results of TC point out to depletion rather than increase, (but these results of TC depends on the choice of the initial galactic helium isotopic ratio (Cameron, private communication))

The cosmological implications were discussed by Geiss and Reeves (1972), Black

(1972), Wagoner (1972), and Reeves *et al.* (1972), in a general nucleosynthetic context. While the elements with $A \geqslant 12$ are most likely of stellar origin, the situation is different of the lighter atoms. An analysis by Meneguzzi *et al.* (1972), following the work of Reeves *et al.* (1970) and Mitler (1971), has shown that while the isotope of ^{6}Li, ^{9}Be, ^{10}B, ^{11}B have most likely originated in the bombardment of the interstellar gas by the galactic cosmic rays, the atoms of H, D, ^{3}He, ^{4}He and ^{7}Li must have another origin. The current belief is that H, D and ^{4}He are child of the Big-Bang. For ^{3}He and ^{7}Li, the situation is more confused (stellar origin remains a possibility) but a Big-Bang origin is also quite promising.

Big-Bang nucleosynthesis is mostly dominated by the choice of two parameters; the baryonic (B) and the leptonic (L_e) numbers of the universe (Fowler, 1971). I have recently tried (Reeves, 1972) to obtain information on these numbers through the fact that we require the Big-Bang to have produced the light elements in quantities which should match the observed values, after appropriate corrections for astraction effects (as discussed previously). For each isotope D, ^{3}He, ^{4}He, ^{7}Li, I have drawn in the Le-B plane the areas corresponding to abundances in agreement with the observations (with wide uncertainties). These areas are then superimposed and the area of coherence are determined according to nucleosynthetic statements of varying credibility.

The more credible statement is that the Big-Bang is responsible for the observed H, D, *and* has not made *more* ^{3}He, ^{4}He, ^{7}Li than observed (it could have made less if these elements were also made elsewhere). From there we obtain $-1 < L_e < 5$ and $10^{-31} < \varrho < 3 \times 10^{-30}$ for the present universal density.

The less credible statement is that all of H, D, ^{3}He, ^{4}He and ^{7}Li were made in the Big-Bang: then we get $L_e \simeq -0.18$ and $\varrho \simeq 2 \times 10^{-30}$ g (about twenty times the density of visible matter and a few times less than the critical density).

Too many uncertainties are attached to the second statement: it should not be taken too seriously at this point. It is however illustrative of a methodology and shows in particular that nucleosynthetic yields is a very sensitive function of the baryonic and leptonic numbers of the universe. As measurements and theories gets better, nucleosynthesis of the light elements should become more and more important in the build-up of cosmological models.

References

Anders, E.: 1972, Proceedings of the Symposium on 'The Origin of the Solar System', Nice, CNRS, 1972.

Beer, R. and Taylor, F.: 1972, to be published.

Beer, R., Farmer, C. B., Norton, R. H., Martonchik, J. V., and Barnes, T. O.: 1972, *Science* **175**, 1360.

Black, D. C.: 1971, *Nature* **234**, 148.

Black, D. C.: 1972, *Geochim. Cosmochim. Acta* **36**, 347.

Black, D. C.: 1972, Proceedings of the Symposium on 'The Origin of the Solar System', Nice, CNRS, 1972.

Boato, G.: 1954, *Geochim. Cosmochim. Acta* **6**, 209.

Craig, H.: 1961, *Science* **133**, 1833.

Danziger, I. J.: 1970, *Ann. Rev. Astron. Astrophys.* **8**, 161.

Fowler, W. A.: 1971, in *The Astrophysical Aspects of the Weak Interactions*, Academia Nazionale de Lincei-Roma, p. 116.

Geiss, J. and Reeves, H.: 1972, *Astron. Astrophys.* **18**, 126.
Geiss, J., Eberhardt, P., Buhler, F., Meister, J., and Signer, P.: 1970, *J. Geophys. Res.* **75**, 5972–5979.
Meneguzzi, M., Audouze, J., and Reeves, H.: 1971, *Astron. Astrophys.* **15**, 337.
Mitler, E. H.: 1970, Smithsonian Institution, preprint No. 004-36.
Predmore, C. G., Goldwire, H. C., Jr., and Walters, G. K.: 1971, *Astrophys. J. Letters* **168**, L125.
Reeves, H.: 1972, *Phys. Rev. D.* **6**, 3363.
Reeves, H. and Bottinga, Y.: 1972, *Nature* **238**, 326.
Reeves, H., Fowler, W. A., and Hoyle, F.: 1970, *Nature* **226**, 727.
Reeves, H., Audouze, J., Fowler, W. A., and Schramm, D. N.: 1973, *Astrophys. J.* **179**, 909.
Truran, J. W. and Cameron, A. G. W.: 1971, *Astrophys. Space Sci.* **14**, 179.
Wagoner, R. V.: 1973, *Astrophys. J.* **179**, 343.
Weinreb, S.: 1962, *Nature* **195**, 367.

Addendum: Recent Developments in 'Deuteronomy' December 1972

Since the summer of 1972 the situation has developed very rapidly on the observational front. There has been a new attempt at detecting the 92 cm line of Deuterium from the Galactic Center, by Cesarsky *et al.* (1972). According to their measurements and interpretation the D/H ratio ranges from : $3 \times 10^{-5} < D/H < 5 \times 10^{-4}$. While the upper limit appears safe, the lower limit is itself based on an evaluation of the upper limit of the hydrogen opacity (in 21 cm) in the same region, which is itself very uncertain.

Second, the molecule DCN has been identified in the direction of the Orion Nebula by Jefferts *et al.* (1973) and also by Wilson *et al.* (1973). Values of DCN/HCN as large as 6×10^{-3} have been reported. The one obvious interpretation of such a large ratio is of course in term of molecular reactions leading to deuterium enrichment in the HCN phase through, for instance:

$$HD + HCN \rightleftharpoons DCN + H_2 .$$

Preliminary investigation by Bottinga (private communication) shows that, as in the case of water and ammonia (Reeves and Bottinga, 1972), the molecular constant become large at low temperature and may well have played a role in an astrophysical scale as for the water in the solar system (Geiss and Reeves, 1972).

The suggestion is even made stronger by the fact that clouds are usually found to be at very low $T(70$ to 125 K$)$. To reconcile the high DCN/HCN in the Orion clouds with the previously stated expected galactic value $(D/H \approx 2 \times 10^{-5})$ we need a molecular constant of ≈ 250 which is well within the realm of the possibilities (an accurate calculation is under way by Bottinga).*

This interpretation takes it for granted that equilibrium will have been reached, which certainly is not guaranteed in these low density clouds. Inversely it would be rather imprudent to determine the temperature from the required value of the molecular constant...

In summary two points come up:

(a) we have now confirmation of the existence of D outside of the solar system. We

---

* An estimation has already been made by Solomon and Woolf (preprint).

know that the total number of deuterium atoms must be large and can not be accounted for by 'local' mechanisms of origin.*

(b) we have no strong reason to reject the view that the present universal abundance of D/H is $\approx 2 \times 10^{-5}$ (Reeves *et al.*, 1973) and that wherever larger ratios are seen in molecular phases, this is due to the effect of molecular exchange reactions in water, methane or cyanidric acid.

Note added in proof. From Lyman lines of D in front of ε Aurigae Rogers and York (1973, report at the IAU Conference of Sydney) have obtained $n(D)/n(H) = (1.5 \pm 0.4) \times 10^{-5}$.

References

Cesarsky, D. A., Moffet, A. T., and Pasachoff, J. M.: 1973, *Astrophys. J. Letters* **180**, 1.
Geiss, J. and Reeves, H.: 1972, *Astron Astrophys.* **18**, 126.
Jefferts, K. B., Penzias, A. A., and Wilson, R. W.: 1973, *Astrophys. J. Letters* **179**, 57.
Reeves, H. and Bottinga, Y.: 1972, *Nature* 238, 326.
Reeves, H., Audouze, J., Fowler, W. A., and Schramm, D. N.: 1973, *Astrophys. J.* **179**, 909.
Wilson, R. W., Penzias, A. A., Jefferts, K. B., and Solomon, P. M.: 1973, *Astrophys. J. Letters* **179**, 107.

* See however Hoyle and Fowler (preprint OAP 305).

THE MOON AS A HIGH TEMPERATURE CONDENSATE* †

DON L. ANDERSON

Seismological Laboratory, Division of Geological and Planetary Sciences,
California Institute of Technology, Pasadena, Calif., U.S.A.

Abstract. The accretion during condensation mechanism, if it occurs during the early over-luminous stage of the Sun, can explain the differences in composition of the terrestrial planets and the Moon. An important factor is the variation of pressure and temperature with distance from the Sun, and in the case of the Moon and captured satellites of other planets, with distance from the median plane. Current estimates of the temperature and pressure in the solar nebula suggest that condensation will not be complete in the vicinity of the terrestrial planets, and that depending on location, iron, magnesium silicates and the volatiles will be at least partially held in the gaseous phase and subject to separation from the dust by solar wind and magnetic effects associated with the transfer of angular momentum just before the Sun joins the Main Sequence.

Many of the properties of the Moon, including the 'enrichment' in Ca, Al, Ti, U, Th, Ba, Sr and the REE and the 'depletion' in Fe, Rb, K, Na and other volatiles can be understood if the Moon represents a high temperature condensate from the solar nebula. Thermodynamic calculations show that Ca, Al and Ti rich compounds condense first in a cooling nebula. The high temperature mineralogy is gehlenite, spinel, perovskite, Ca-Al-rich pyroxenes and anorthite. The model is consistent with extensive early melting, shallow melting at 3 AE and with presently high deep internal temperatures. It is predicted that the outer 250 km is rich in plagioclase and FeO. The low iron content of the interior in this model raises the interior temperatures estimated from electrical conductivity by some 800 °C. The lunar crust is 80% gabbroic anorthosite, 20% basalt and is about 250–270 km thick. The lunar mantle is probably composed of spinel, merwinite and diopside with a density of 3.4 g cm⁻³.

1. Introduction

The inhomogeneous accretion hypothesis has recently been revived by Clark, Turekian and Grossman (1972) to explain the gross layering in the Earth and the differences in composition of the terrestrial planets. Larimer (1967) and Grossman (1972) computed the condensation sequence of elements and compounds in a cooling solar nebula and have applied these calculations to the problem of differences in the composition of meteorites. Earlier, Hoyle and Wickramasinghe (1968) proposed that the solid material in the solar system condensed and accreted during the early over-luminous phase of the contracting Sun, during the short period of time that the Sun was rotationally unstable and transferring angular momentum to the planetray material, and before the slight increase in luminosity that occurred just before the Sun joined the main sequence. They revived the old idea that the planets formed hot. Most current theories of the origin of the planets assume that they accreted cold from either a mixture of iron and silicates or from material of chondritic composition which was chemically reduced in the terminal stages of accretion to yield free iron. In

* Paper dedicated to Prof. Harold C. Urey on the occasion of his 80th birthday on 29 April 1973.
† Contribution No. 2260, Division of Geological and Planetary Sciences California Institute of Technology, Pasadena, Calif. 91109, U.S.A. Presented at the *IAU Symp. Cosmochem.*, Cambridge, Mass. August 14–16, 1972.

the former case radioactive heating is invoked to melt and separate the Earth's molten core. In the latter case, the gravitational potential energy of the accreting planet is used both to reduce the iron and to melt it so it can drain to the interior. There are serious time scale and compositional difficulties with both of these hypotheses. The accretion during condensation hypothesis, while the Sun is still in the over-luminous stage, can avoid these difficulties and can explain the differences in the bulk composition of the planets. It also provides a clue to the anomalous situation of the Moon.

In particular, if it is assumed that condensation and accretion occurred before the dissipation of the solar nebula and that the nebula contained uncondensed silicates and volatiles in the vicinity of the terrestrial planets, it is fairly easy to understand the differences in bulk chemistry of the terrestrial planets, the Moon and the meteorites. The bulk chemistry of the various bodies will reflect the temperatures prevailing in their vicinity just prior to dissipation, in particular, the composition of the material that has condensed at this temperature.

TABLE I

Stability Fields of equilibrium condensates
at 10^{-3} atmospheres total pressure

Phase		Condensation temperature (K)
Trace refractories (1)		1931–1768
Corundum	Al_2O_3	1758
	HfO_2	1744
	Mo	1698
Perovskite	$CaTiO_3$	1647
	Ru	1634
Melilite		
Gehlenite	$Ca_2Al_2SiO_7$	1625
Akermanite	$Ca_2MgSi_2O_7$	
	ThO_2	1517
Spinel	$MgAl_2O_4$	1513
Merwinite	$Ca_3MgSi_2O_8$	1475
Metallic Iron	(Fe, Ni)	1473 (2)
Diopside	$CaMgSi_2O_6$	1450
Forsterite	Mg_2SiO_4	1444
	Ti_3O_5	1393
Anorthite	$CaAl_2Si_2O_8$	1362
Enstatite	$MgSiO_3$	1349
Eskolaite	Cr_2O_3	1294
Rutile	TiO_2	1125

(1) Os, Sc_2O_3, Te, Ta, ZrO_2, W, Nb, Y_2O_3 etc.
(2) The relative location of Fe–Ni in the condensation sequence depends critically on pressures in the nebula and on departures from equilibrium, i.e , the nebula may be supersaturated in iron vapor before condensation ensues. Iron condenses after forsterite, and presumably diopside, at pressures less than 10^{-4} atm.

2. Condensation

The condensation sequence of elements and compounds from a cooling cloud of solar composition has been calculated by Larimer (1967), Lord (1965), and Grossman (1972). The early condensates are Al, Ca and Ti compounds such as gehlenite (Ca_2Al_2-SiO_7), spinel ($MgAl_2O_4$) and perovskite ($CaTiO_3$). These compounds all condense before iron. The relative absence of iron in the Moon suggests that it may have accreted from these compounds. Under non-equilibrium or lower pressure conditions such compounds as diopside ($CaMgSi_2O_6$), forsterite (Mg_2SiO_4) and anorthite ($CaAl_2Si_2O_8$) also condense before iron. The early condensate will be enriched in the REE and other refractories which substitute readily for Ca, and may be enriched in Th and U which are relatively refractory. K_2O, S, Na_2O, H_2O and other volatiles will be deficient in the early condensate.

The condensation sequence of the compounds and elements in a cloud of solar composition is shown in Table I. The Ca, Al and Ti rich compounds will provide the nucleus for later condensates such as iron and the magnesium-rich silicates. We will refer to the Ca, Al and Ti rich compounds as the early condensates or the refractories.

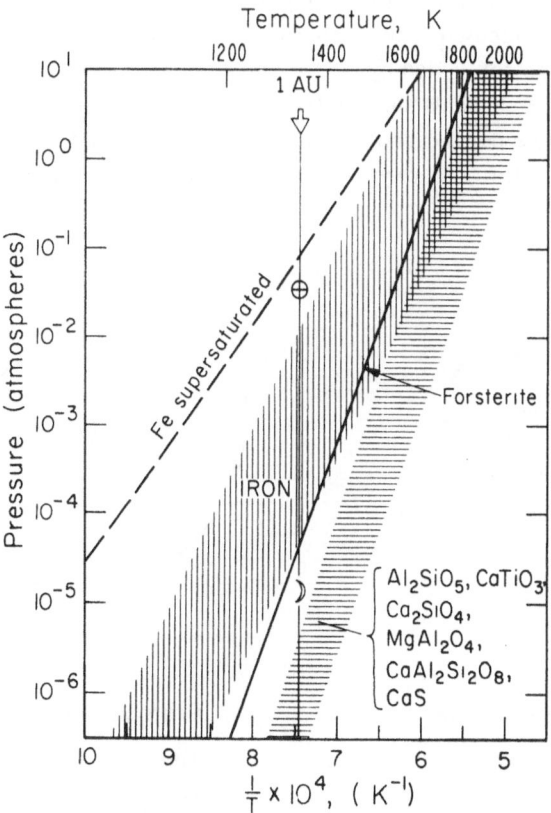

Fig. 1.   Condensation temperature vs pressure in the nebula.

Figure 1 gives the condensation temperature as a function of pressure in the nebula and also the temperature in the vicinity of the Earth during the over-luminous phase of the Sun. As drawn, the Earth will be composed of the early condensates plus iron and the magnesium silicates in solar or cosmic proportions. The placement of the Moon anticipates our later conclusion that the Moon is composed of refractories and consists of material that condensed at different pressures, and therefore, a different time, than the bulk of the Earth. The other terrestrial planets can also be placed on this diagram. For example, Mercury will be in a higher pressure and higher temperature part of the nebula. In order to explain its high mean density we can assume that it is composed primarily of the refractories and iron and that it has not received a full complement of magnesium silicates such as olivine and pyroxene. Venus is slightly lighter than the Earth and, therefore, probably did not incorporate as much iron into its interior. This suggests that the nebula was dissipated before iron had finished condensing in the vicinity of Venus. Mars was in a much cooler part of the nebula and incorporated volatiles as well as the refractories, iron and the magnesium silicates into its interior. The carbonaceous chondrites are presumably even more volatile rich than Mars if they formed beyond the orbit of Mars.

Figures 2 and 3 indicate in a schematic fashion how the composition of a planet changes as the temperature in the nebula decreases. At high temperatures only the refractories have condensed. The composition of the solid material in the solar system at this point, will be dominated by Ca, Al and Ti rich minerals and will be enriched in the refractory trace elements such as Ba, Sr, Th and the rare Earth elements. Inclusions in type II and type III carbonaceous chondrites have these characteristics and, in addition, have textures and oxygen isotope ratios that indicate that they are high

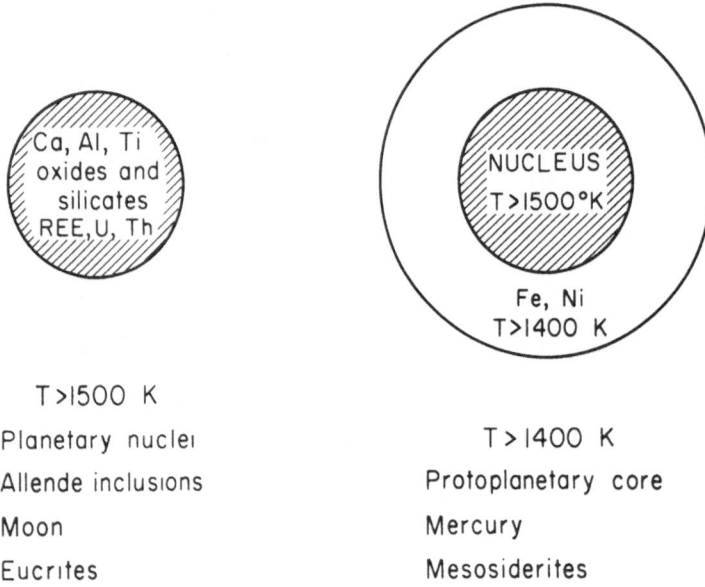

Fig. 2.   Development of a planet according to the accretion during condensation hypothesis.

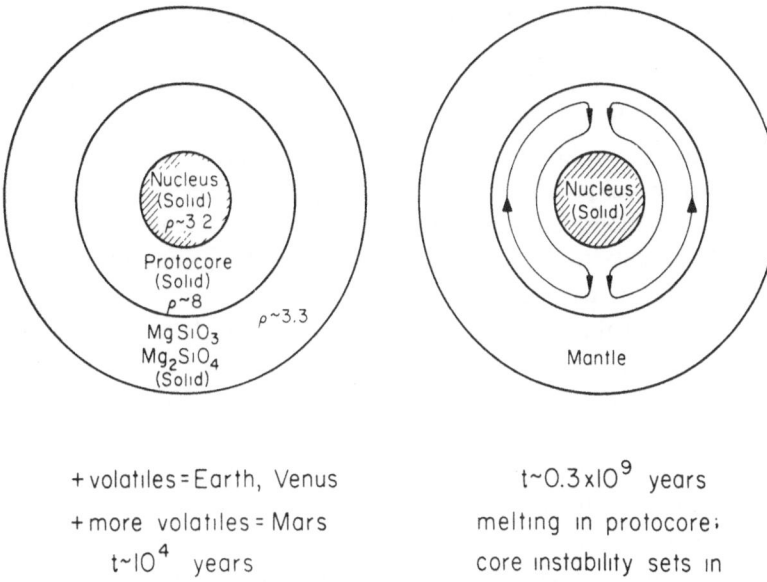

+ volatiles = Earth, Venus $t \sim 0.3 \times 10^9$ years

+ more volatiles = Mars   melting in protocore;

$t \sim 10^4$ years   core instability sets in

Fig. 3.   Development of a planet, continued.

temperature condensates. The Moon also has chemical and physical properties that indicate it is composed of the high temperature condensates.

Iron condenses at lower temperatures and, along with the refractories, forms the protocore of a planet. Mercury may represent a planet that has not proceeded much beyond this stage. The bulk of Venus, Earth, and Mars are composed of the ferro-magnesium silicates which condense during and after iron. Minerals such as olivine, pyroxene and garnet form the major part of the mantle of a fully assembled terrestrial planet. Prior to the condensation of iron the mineralogy of a planet is dominated by such exotic minerals as perovskite, akermanite, gehlenite, spinel and merwinite.

3. The Moon

The low iron content of the Moon, compared to terrestrial, solar or meteoritic abundances, has lead to many discussions of metal-silicate fractionation mechanisms in the solar nebula and has been used as an argument for both a fission and a capture origin for the Moon. The Surveyor and Apollo missions have shown that the composition of the Moon is anomalous on other counts. It is depleted in volatiles, as well as iron, and is enriched in refractories. This is true not only for the surface rocks but for their source regions as well and, therefore, applies to a considerable fraction of the lunar interior. Therefore, it is clear that more than just metal-silicate fractionation is required in order to create a Moon from solar or 'cosmic' abundances. Although many and diverse proposals have been put forth to explain the bulk and surface chemical properties of the Moon, most of them assume that material of chondritic composition was important sometime in the Moon's ancestry.

We put forth the alternate hypothesis that the bulk of the Moon is composed of those elements and compounds that condensed prior to the condensation of iron. Iron, $MgSiO_3$, Mg_2SiO_4 and the volatiles were incorporated into the interior in only minor amounts, and probably, only during the terminal stages of accretion. The outer part, ~250 km, of the Moon in our model is almost identical to that proposed by Gast (1972) on geochemical grounds. However, the deep interior is CaO and Al_2O_3 rich and is dominantly diopside, merwinite and spinel. This assemblage has acceptable densities and is stable to higher pressures than the Ringwood-Essene (1970) low Ca–Al model lunar pyroxenite.

The enrichment of the Moon in refractories and its depletion in volatiles is now well documented. Figure 4 gives lunar vs meteoritic abundances. Note that carbonaceous chondrites are a very poor approximation to the composition of the lunar surface rocks. The lunar rocks are fairly uniformly enriched in refractories and strongly depleted in volatiles. However, the Ca–Al rich inclusions in the Allende meteorite provide an excellent match to the lunar surface composition.

The early condensate amounts to 5.8–9.0% of the total and the enrichment factor of the trace element refractories is 11–17 times chondritic. In this regard it is of interest that the refractory trace elements are enriched in the Allende inclusions by a factor of

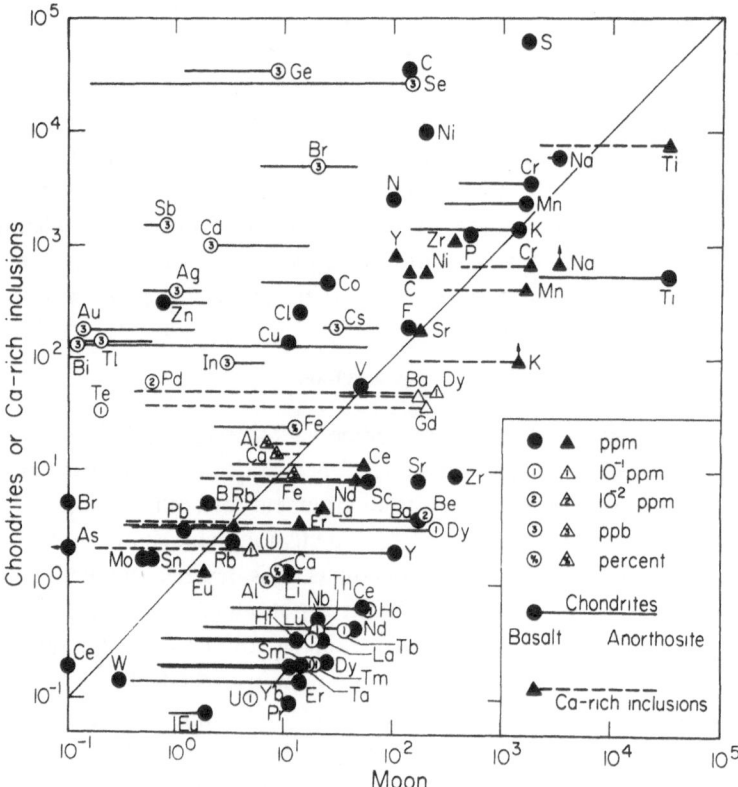

Fig. 4.   Lunar vs meteoritic abundances.

11 over the whole meteorite and that these inclusions comprise ~8% of the meteorite, a C3 chondrite.

Figure 5 shows lunar abundances, normalized to carbonaceous chondrites, plotted as a function of condensation temperature. The amount of material that has condensed and the predicted enrichment factor of the trace element refractories is also shown. Note that the lunar surface material is enriched by about the predicted amount

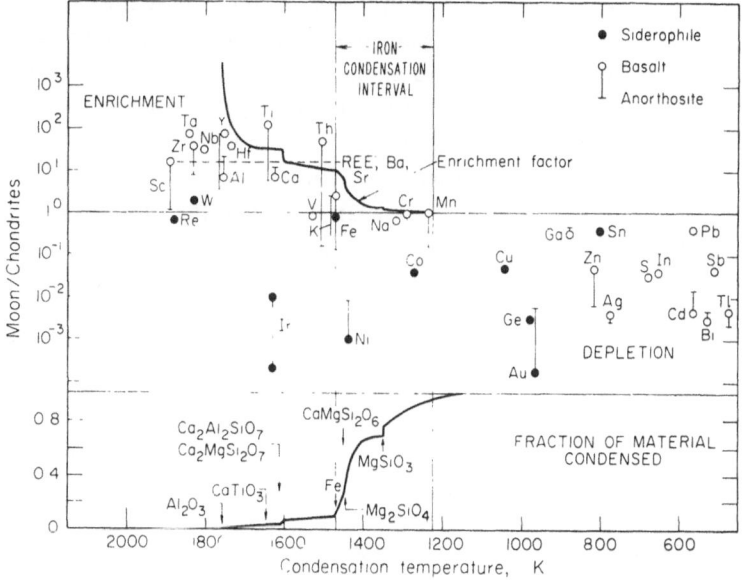

Fig. 5.   Chondritic normalized lunar abundances (above) and fraction of condensed material (below) as a function of temperature in the nebula.

for the pre-iron condensate. Note also the very rapid increase in the condensed fraction that occurs when iron and olivine start to condense. Conditions in the solar nebula, such as opacity and cooling rate, can be expected to change markedly as temperatures drop to this level.

The condensation temperatures of the REE, Ba and Sr are uncertain but their average enrichment, shown by the dashed line, is about the same as the other refractories suggesting that they condense over a similar temperature range. Grossman (1972) calculates that U, La, Sm and Eu will condense below 1473 K at 10^{-3} atm. The U/Th ratio, therefore, may be lower in the early condensates than in chondrites. It is likely, however, that solid solution effects will allow the condensation of some refractory trace elements at higher temperature than is the case for the pure phases. Perovskite, for instance, could provide lattice sites for the removal of the rare Earth and other trace elements at temperatures of 1647 K. The rare Earth elements will probably condense over the same temperature interval as corundum, perovskite, melilite and spinel. *Ir* is the main anomaly in Figure 5. It is commonly concentrated in residual high temperature crystals such as chromite and spinel. It may, therefore, reside in a dense phase in the interior of the Moon or, because it is siderophile, it may never have entered the Moon in quantity.

The refractory properties of the Moon apply not only to the surface rocks, but also to the inferred source region of these rocks, which extends at least 100 km and possibly 400 km into the interior. This has been eloquently demonstrated by Paul Gast and his coworkers. In any event, a substantial fraction of the Moon seems to be enriched in Ca and Al and the trace refractories such as Ba, Sr and the REE. Of course, the mean density and the moment of inertia of the Moon require that the whole Moon be deficient in iron relative to solar, terrestrial or chondritic abundances. We have previously suggested that the whole Moon may also be deficient in those elements and compounds that condense after iron. It has been suggested by many authors, however, that the deep interior of the Moon may be chondritic in composition. This suggestion is based on the argument of Ringwood and Essene (1970) that both CaO and Al_2O_3 must be less than about 6% if the mean density of the Moon is to be satisfied. This constraint is not valid and there is no support for the suggestion that the interior of the Moon must be more volatile-rich and less refractory than the outer shell.

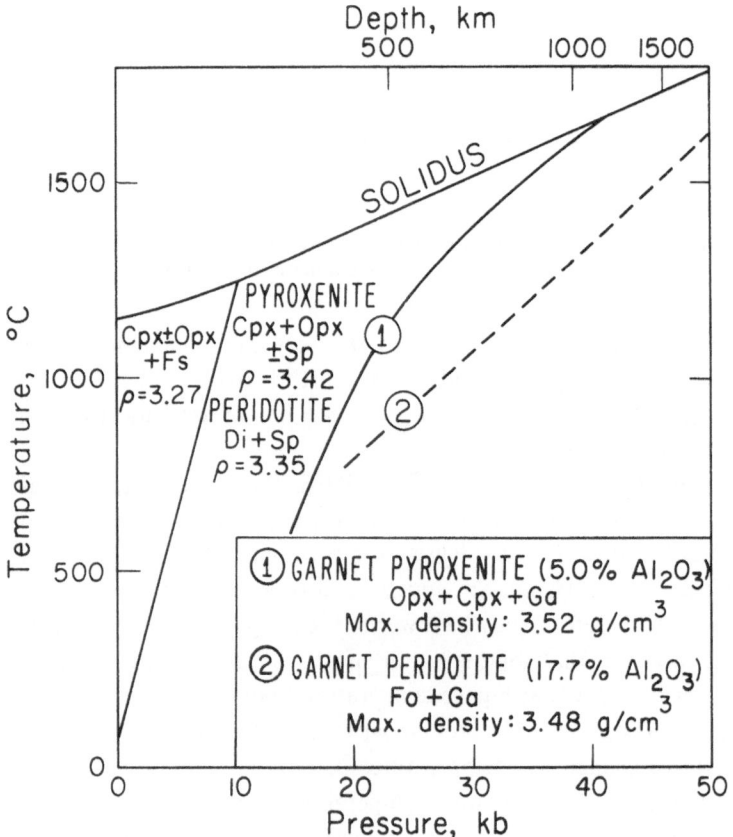

Fig. 6.   Stability fields for two lunar interior models (Anderson, 1972d).

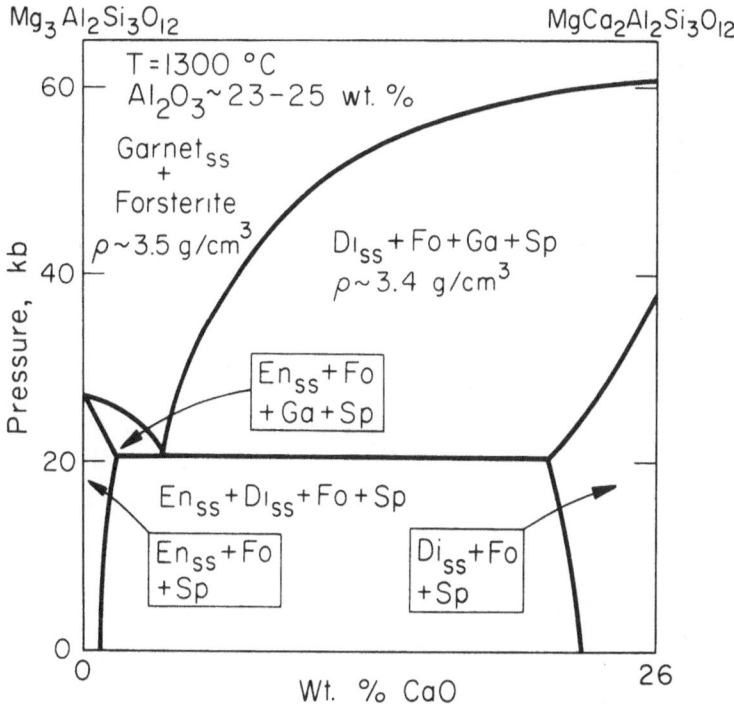

Fig. 7.   Phase relations in a simplified periotite (Anderson, 1972d).

For example, Figure 6 gives the stability fields of a $CaO-Al_2O_3$ rich peridotite and the 'model lunar pyroxenite' of Ringwood and Essene (1970). Note that the intermediate density field for the peridotite is broader than the corresponding field for the $CaO-Al_2O_3$ poor pyroxenite and the density in this field closely corresponds to the mean density of the Moon. The composition of this peridotite is close to that of the composition of the early condensate with the low melting fraction removed. To put it another way the composition of this material is similar to the near liquidus crystals of the Allende inclusions.

Figure 7 gives the equilibrium fields in a simplified peridotite as a function of CaO content. At low $CaO-Al_2O_3$ contents the high density garnet rich assemblage occurs at shallower depths as the CaO content increases. However, at high CaO contents the intermediate density field broadens rapidly. Therefore, there is no difficulty in having the bulk of the Moon composed of the Ca–Al rich refractories.

Likewise, a thick $CaO-Al_2O_3$ and plagioclase rich outer shell is permitted by the mean density of the Moon. Figure 8 shows the stability fields of several high Ca–Al plagioclase bearing assemblages. The intermediate density field is stable throughout most of the lunar crust and mantle. Increasing the Al_2O_3 content, in fact, increases the pressure of the gabbro-eclogite phase change, Figure 9.

All of the geophysical and geochemical data suggest, or at least are consistent with, the idea that the bulk of the Moon is composed of material that condensed from the solar nebula prior to the condensation of appreciable amounts of iron.

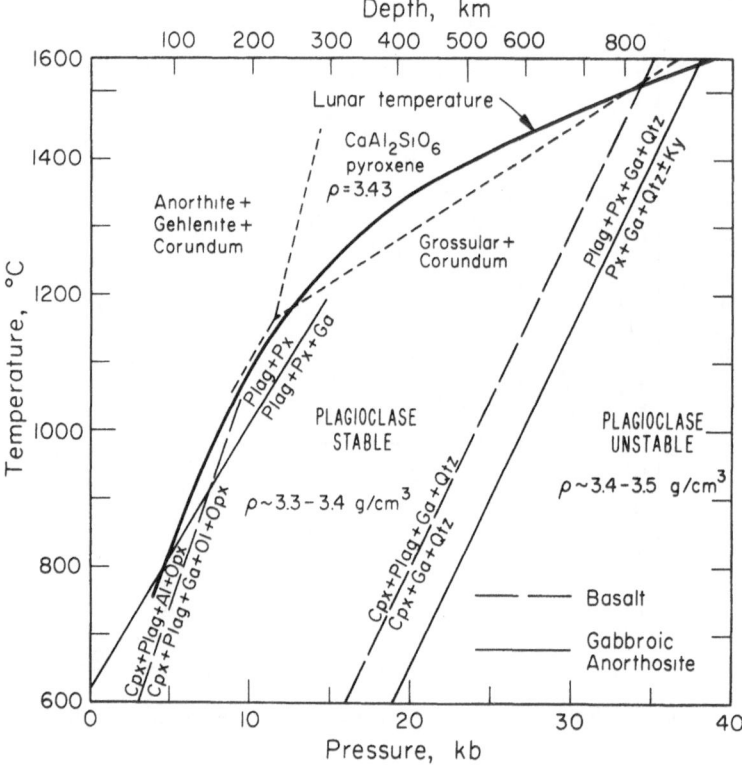

Fig. 8.   Stability fields of plagioclase in plagioclase rich systems (Anderson, 1972d).

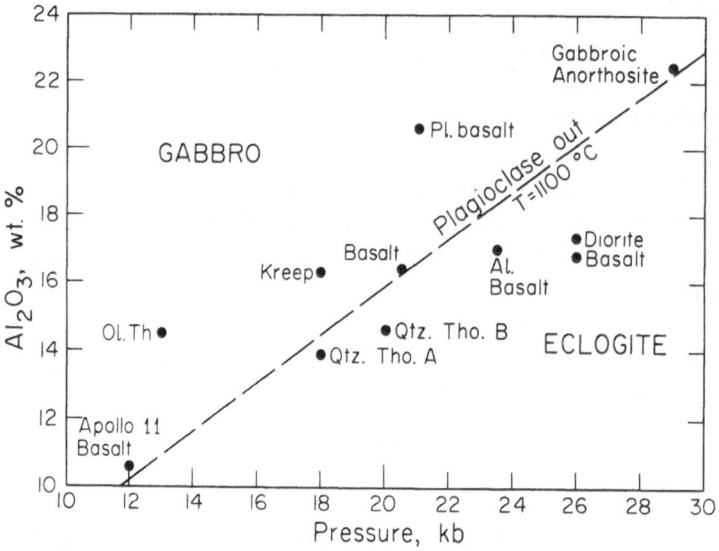

Fig. 9.   Gabbro-eclogite boundary as a function of Al₂O₃ content (Anderson, 1972d).

The following is one possible scenario for the formation and subsequent evolution of the Moon: The Moon started to accrete during or shortly after the initiation of condensation in the solar nebula. Temperatures in its vicinity never fell far below the condensation temperature of iron before the uncondensed portion of the solar nebula was removed. It is, therefore, composed of corundum, perovskite, melilite (a solid solution of gehlenite and akermanite), spinel and possibly, diopside and anorthite. Its composition is similar to the Ca–Al rich inclusions in the Allende meteorite, and it is enriched in the trace element refractories such as Ba, Sr, REE, U and Th. Because of the high initial temperatures (i.e., condensation temperature), the rapid accretion rate and the high content of U and Th, it melts, either partially or totally, during or shortly after accretion. The near liquidus crystals, spinel, melilite and perovskite, which represent about 65% of the early condensate settle to the interior. The remainder constitutes the outer 250 km of the Moon and is the source region for the lunar basalts and anorthosites. Both the interior and the outer shell are enriched in Ca, Al and in Ti. The outer shell will be enriched in U, Th, Fe, and Ti which are rejected by the early crystallizing solids. Anorthite crystallizes from the residual melt before pyroxene and, being lighter than the remaining melt, floats to the surface to form the lunar

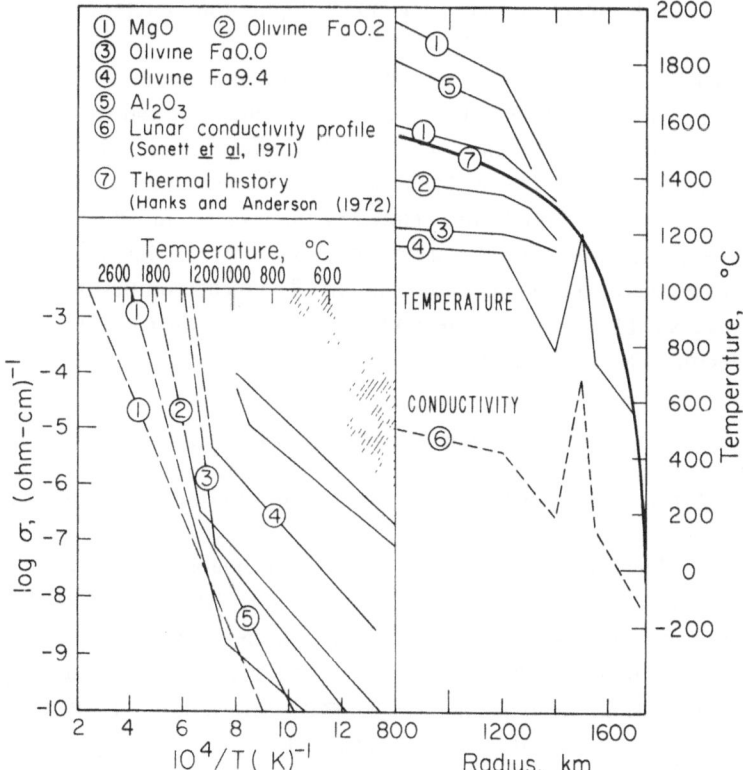

Fig. 10. Electrical conductivities of MgO, Al₂O₃, forsterite (Mg₂SiO₄) and olivines of varying fayalite (Fe₂SiO₄) content (left). Electrical conductivity of the Moon (lower right) and inferred temperatures (upper right).

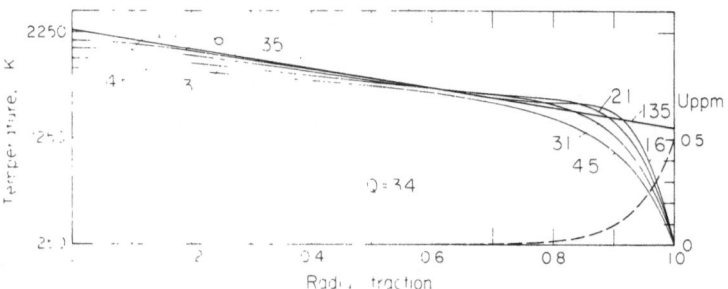

Fig. 11.   Thermal history calculations (Hanks and Anderson, 1972).

protocrust, in particular, the highlands. The early condensate has the potential to yield a layer about 200 km thick of anorthosite. The lunar basalts are the result of further fractional crystallization or later partial melting processes. The basalts represent about 20% of the outer shell. With this model the outer 250 km contains most of the iron of the Moon and the interior is essentially iron-free. This requires a reinterpretation of the lunar electrical conductivity profiles. Figure 10 gives the lunar conductivities as inferred by Sonett *et al.* (1971) and temperatures inferred from iron poor refractories. Temperatures are some 800 °C higher than inferred for an iron rich interior.

Figure 11 gives thermal history calculations which indicate, in agreement with the previous figure, that the interior of the Moon is presently hot and was hot enough in its early history to supply basalts for some 10^9 yr after initial melting from depths greater than about 150 km. The thickness of the solid 'lithosphere' increases with time.

With this model the lunar interior, below 250 km, is merwinite, diopside, spinel and perovskite with a zero-pressure density of about 3.4 g cm^{-3}. Merwinite and diopside are the high pressure equivalents of akermanite. The mean composition of the outer shell or crust, is gabbroic anorthosite, and it is enriched in the trace element refractories by a factor of about 16 times chondritic.

4. The Allende Meteorite

Theoretically, the early condensates from a cooling solar nebula include perovskite, spinel, gehlenite, diopside, anorthite and other Ca, Al and Ti compounds. This assemblage will be enriched in such refractory trace elements as REE, Sr, and Ba and possibly Th and U, and depleted in such volatiles as K, Rb, S and H_2O. Type II and III carbonaceous chondrites contain Ca–Al rich inclusions which involve these minerals and which are enriched in Ba, Sr and the REE to about the extent predicted in the previous section. The bulk chemistry of these inclusions in the Allende meteorite is high in Ca, Al and Ti and low in Fe, Mg and volatiles. The dominant minerals include gehlenite ($Ca_2Al_2SiO_7$), spinel ($MgAl_2O_4$), fassaite, an aluminous titanium-rich pyroxene ($Ca(Mg, Al, Ti)(Al, Si)_2O_6$) and anorthite ($CaAl_2Si_2O_8$). Other minerals include perovskite ($CaTiO_3$), diopside ($CaMgSi_2O_6$), ferroaugite ($Ca(Fe,$

Mg, Al)(Al, Si)$_2$O$_6$), grossular (Ca$_3$Al$_2$Si$_3$O$_{12}$) and corundum (Al$_2$O$_3$). The inclusions contain no metallic iron or olivine. These aggregates have been studied in detail by Clarke *et al.* (1970), Marvin *et al.* (1970), Gast *et al.* (1970), and Grossman (1972). The texture and the presence of reaction rims indicate that the inclusions were inserted into the matrix at high temperature. The δO^{18} value of this material is much more negative than any other meteoritic material (Onuma *et al.*, 1972) and in the range to be expected for a primary high-temperature condensate from a nebular gas. Marvin *et al.* (1970) and Clarke *et al.* (1970) have noted the similarity between the composition and mineralogy of the aggregates and the early condensates in a cooling nebula. It is extremely unlikely that the peculiarities of these inclusions could be a result of igneous differentiation processes acting on material of solar or chondritic composition. On the basis of their bulk chemistry, mineralogy, texture and oxygen isotope ratios, the Ca–Al rich inclusions apparently represent the highest temperature condensates from a gas of solar composition and are, therefore, the most primitive solids in the solar system. Grossman (1972) has supported this view with detailed thermodynamic calculations. If this interpretation is correct they are extremely significant in understanding the origin and composition of the Moon. We will show that many properties of the Moon, including its overall physical properties, gross layering and the properties of the source region of lunar igneous rocks can be explained if the bulk composition of the Moon is similar to that of the Ca–Al rich inclusions.

5. Trace Elements

The concentrations of the trace element refractories have been determined for the Ca–Al inclusions of the Allende chondrite by Gast *et al.* (1970) and Grossman (1972) and for the whole meteorite by Clarke *et al.* (1970) and Wakita and Schmitt (1970a, b). The average enrichment of these elements in the inclusions, which make up ~8% of the meteorite, is about a factor of 11, implying that these refractory elements are almost entirely contained in the inclusions, and, by implication in the earliest condensates.

Table II gives concentrations of Ba, La, REE, Sr, Rb, K and U for the Ca–Al rich inclusions of the Allende meteorite, the whole meteorite, carbonaceous chondrites, Apollo 11 basalts and a lunar anorthosite. The abundances in the inclusions are approximately 16 times the chondritic abundances, column (3). In this respect the Ca–Al inclusions are a much more satisfactory source for the lunar igneous material than are carbonaceous chondrites. The first column are abundances in an Allende Ca–Al rich inclusion; the second column are whole meteorite abundances. The anorthosite and basalt have large and opposite europium anomalies. The anomalies can be suppressed by mixing basalt with about five times as much anorthosite, columns (6) and (7). This is also shown in Figure 12. The mixture is normalized to carbonaceous chondrites in column (7) and to the Allende inclusions in column (8). The lunar basalts are enriched in refractories by more than an order of magnitude relative to carbonaceous chondrites and a factor of five relative to the Allende inclu-

TABLE II

Trace elements in allende meteorite, carbonaceous chondrites and the Moon

	Meteorites			Moon				Moon/Meteorite	
	(1)	(2)	(3)	(4)	(5)	(6)	(7)	(8)	(9)
Refractories									
Ba	47.3	5	3.6	200	6.28	48.9	33.4	13.6	1.03
La	4.63	0 44	(0.28)	18	0.12	4.5	2.62	14.5	0.87
Ce	11.5	1.25	0.787	54	0.35	12.2	7.86	15.5	1.06
Nd	8.40	0.91	0.652	46	0.18	10.3	6.59	15.8	1.23
Sm	2.82	0.29	0.208	15	0.05	3.4	2.14	16.2	1.21
Eu	1.30	0.11	0.071	2	0.81	1.1	0.96	14.8	0.81
Gd	3.87	0.43	0.256	20	0.05	4.4	2.84	17.2	1.14
Dy	4.90	0.42	0.303	25	0.04	5.5	3.53	18.2	1.12
Er	3.44	0.31	0.182	14	0.02	3.1	1.97	17.0	0.90
Yb	3.96	0.32	0.188	13	0.04	2.9	1.85	15.4	0.73
Sm/Eu	2.17	2 64	2.93	7.89	0.06	3.24			
U	(0.2)	0.019	0.01	0.5	0.015	0.12	0.083	12.	0.60–
	0.03								4.00
Sr	180	13	11	170	178	176	177	16.0	0.98
Volatiles									
Rb	3.5	1.3	3.0	3.4	0.15	0.87	.61	0.29	0.24
K	96–415	250	1000	1400	120	402	299	0.40	4.17–
									0.97
K/U	500–10000	1.3×10^4	10^5	2800	9500	3350			
K/Ba	2–9	50	278	7	19	8			
K/Rb	30–120	192	330	412	800	462			
Rb/Sr	0.019	0.10	3.67	0.02	0.0008	0.0049			

(1) Allende Ca-rich inclusions (Gast et al., 1970); U = (0.2) estimated from 10 × whole meteorite and 16 × C1 chondrites; U = 0.03 from Grossman (1972).
(2) Allende – whole meteorite (Clarke et al., 1970).
(3) Carbonaceous chondrites.
(4) Apollo 11 basalt-mean (Mason and Melson, 1970).
(5) Lunar anorthosite – 15415. 11 (Hubbard et al., 1971).
(6) 0.22 basalt +0.78 anorthosite.
(7) 0.14 basalt + 0.86 anorthosite.
(8) Column 6 normalized to carbonaceous chondrites.
(9) Column 6 normalized to Allende Ca-rich inclusions.

sions. If the composition of the outer shell of the Moon can be accounted for entirely by a mixture of basalt plus anorthosite, the absolute abundances can be made comparable to Allende inclusions as shown in column (9). This mixture, however, still has a small europium anomaly relative to either carbonaceous chondrites or the Ca–Al inclusions.

Table III gives the chondritic normalized trace element refractory abundances for the Allende inclusion and several combinations of the lunar surface material. We obtain, column (1), an enrichment factor of 16. Column (2) is the mixture of basalt and anorthosite required to achieve Allende abundance levels. Column (3) gives the mixture of anorthosite and basalt required to satisfy the Allende Sm/Eu and Eu/Gd

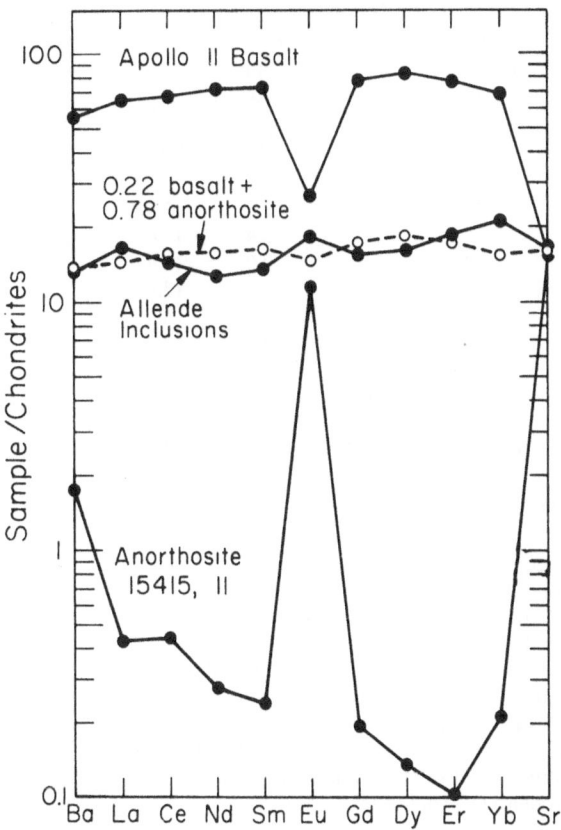

Fig. 12. Trace element refractory abundances in the Moon and the Allende meteorite
(Anderson, 1972c).

ratio. Column (4) is the mixture required to achieve the carbonaceous chondritic
ratios. In all cases appreciable Sr, Ba and the REE must be retained in the interior if the
Moon has the abundances of the Ca–Al rich inclusions. Since the residual crystals in
our model amount to ~65% of the mass of the Moon, the average REE concentration
in the interior is 17–19 times chondritic levels. The Eu anomaly, relative to the Ca–Al
inclusions, can be eliminated by mixing 0.86 anorthosite and 0.14 basalt, column (9).
The exact proportions depend on the choice of materials, but Wakita and Schmitt
(1970) obtained almost ʼdentical values. The lunar interior, for our model (spinel,
melilite and perovskite) however, is quite different from those assumed by the above
authors. These high temperature crystals are able to retain the large ions much more
efficiently than olivines and pyroxenes and these ions will not be as effectively concentrated in the melt. Tables II and III give strong support to the hypothesis that the
lunar differentiates involve a primitive, refractory source region and are consistant
with Gast's (1972) conclusions regarding the nature of the source region.

The refractory trace elements support the hypothesis that the Allende inclusions represent the early condensates of the cooling solar nebulae. If the refractory trace

TABLE III

Enrichment of refractories
Allende inclusions/carbonaceous chondrites and
Lunar surface/carbonaceous chondrites

	(1)	(2)	(3)	(4)
Ba	13.1	13.6	9.3	12.1
La	16.5	14.5	9.4	12.6
Ce	14.6	15.5	10.0	13.4
Nd	12.9	15 8	10 1	13.6
Sm	13.6	16.3	10.3	13.9
Eu	18.3	14.8	13.5	14.3
Gd	15.1	17.2	11.1	15.0
Dy	16.2	18.2	11.7	15.8
Er	18.9	17.0	10.8	14.7
Yb	21.1	15.4	9.8	13.3
Sr	16.4	16.0	16.1	16.0
Average	16.1	15.8	11.1	14.1

(1) (Allende Ca–Al rich inclusions)/carbonaceous
    chondrites.
(2) (0.22 basalt + 0.78 anorthosite)/carbonaceous
    chondrite
(3) (0.14 basalt + 0.86 anorthosite) (to eliminate Eu
    anomaly relative to Allende inclusion)/ carbo-
    naceous chondrites
(4) (0.19 basalt + 0.81 anorthosite) (to eliminate Eu
    anomaly relative to carbonaceous chondrite)/
    carbonaceous chondrites

elements such as Ba, Sr and the REE condense early they will be concentrated in the early condensate relative to their concentration in the Sun, or relative to material such as carbonaceous chondrites, which are presumably representative of the bulk composition of the nebulae. Previously, we estimated that the early condensates would be enriched by a factor of 11–17 in refractory trace elements, relative to solar or chondritic abundances.

6. Major Elements

Table IV gives the major oxide composition of lunar surface material and the Ca–Al rich inclusion, and for comparison, a theoretical estimate of the composition of the early condensate. Columns (3) and (4) are two estimates of the mean composition of the lunar crust based on trace element concentrations. Column (3) is the basalt-anorthosite mixture which is required to give Allende inclusion trace element refractory levels. Column (4) is the mixture which gives the Allende inclusion Sm/Eu ratio. There is little difference as far as the major elements are concerned. In the CaO–MgO––Al_2O_3–SiO_2 system with this composition, pyroxene melts at 1235 °C, anorthite at 1250 °C and gehlenite at about 1400 °C. Spinel remains as a solid until 1550 °C. The first melt will, therefore, be rich in pyroxene and as melting proceeds, will become

TABLE IV

Composition of lunar and Allende materials

	(1)	(2)	(3)	(4)	(5)	(6)	(7)	(8)
SiO₂	40.4	45.7	44.5	45.0	41.4	33.7	30.8	29.5
Al₂O₃	9.4	30.6	25.9	27.6	25.5	26.6	27.6	30.0
FeO	19.3	4.5	7.7	6.6	7.5	2.3	0	–
MgO	7.2	4.8	5.3	5.1	3.3	13.1	17.8	19.7
CaO	11.1	15.8	14.7	15.1	21.3	21.6	22.2	20.7
TiO₂	10.9	0.2	2.5	1.7	0.8	1.3	1.6	–

(1) Apollo 11 basalt-mean.
(2) Lunar anorthosite.
(3) 0.22 basalt +0.78 anorthosite (based on trace elements).
(4) 0.14 basalt +0.86 anorthosite (based on trace elements).
(5) Low melting fraction of Allende inclusions; pyroxene + anorthite (40%)
(6) Allende Ca–Al rich inclusion.
(7) Allende inclusion with low melting fraction removed – i.e. implied composition of the lunar interior if the Moon is composed of the high temperature condensates.
(8) Composition of early condensate ($T > 1450$ K, $P_T = 10^{-3}$ atm) (Grossman, 1972).

more anorthositic. Column (5) gives the composition of the early $T < 1250°C$, melt and also the late condensate and can be compared with columns (3) and (4). The amount of this material in the inclusion corresponds to a thickness of ~ 270 km on the Moon.

Column (6) gives the composition of the inclusion as determined by Clarke *et al.* (1970). Column (7) gives the composition of the high temperature crystals in the Allende inclusion, a possible composition of the deep, >270 km, interior. It would perhaps be surprising if the single Allende aggregate that has been analyzed were completely representative of the early condensate. An alternate approach is to consider the mineralogy of the early condensate predicted from thermodynamic calculations. Column (8) gives the composition of the condensate prior to the condensation of iron, olivine and enstatite. The major difference between the theoretical composition and the Allende aggregate composition is the MgO content. This is probably because the Allende inclusion was armored from complete reaction with $MgO(g)$, therefore keeping the akermanite content of the melilite below equilibrium levels.

It appears that the early condensate is capable, in principal, of satisfying the major element and trace element requirements placed on the interior by the lunar igneous rocks. It is also capable of explaining the geophysical data.

7. Possible Fractionation of an Allende-Like Moon

The composition of the Allende inclusion, as reported by Clarke *et al* (1970) has been recast into a mineral assemblage with the following results (in weight percent); feldspar 28.4%, melilite 39.7%, spinel 25.7%, perovskite 2.3% and diopside 3.9%. The near liquidus phases in a similar assemblage studied by Prince (1954) are spinel and melilite. The crystalline sequence is spinel at about 1550°C followed by melilite at 1400°C, anorthite at 1250°C followed shortly by pyroxene at 1235°C. Fractional crystallization would give a spinel-melilite or spinel-merwinite-diopside interior and a feldspathic pyroxenitic surface layer. The refractory interior would comprise approximately 65% of the mass of the Moon. The residual liquid would be enriched in Fe and any trace elements that are incompatible with the spinel and melilite lattices and would comprise the outer 250 km of the Moon. The basalts and anorthosites could be derived from this layer either by further crystal fractionation involving plagioclase floatation, or by partial melting after solidification. The high U and Th contents of the surface layer are adequate to remelt the lower portions within several hundred million years after solidification (Hanks and Anderson, 1972).

TABLE V

Model compositions of lunar interior

Crust					Mantle	
	(1)	(2)	(3)	(4)	(5)	(6)
SiO_2	48.8	47.7	47.1	52.9	26.7	61.0
TiO_2	3.8	2.7	3.2	0.9	–	0.6
Al_2O_3	25.3	27.8	27.3	17.6	28.3	2.4
MgO	2.1	5.7	6.8	15.4	19.7	24.3
CaO	16.6	15.8	15.6	13.2	25.2	6.5

(1) Allende inclusion with near liquidus phases removed (spinel and melilite); this is the inferred parent liquid for the lunar basalts and anorthosites (this paper).
(2) Average crustal composition derived by mixing basalt and anorthosite in the proportions 22% basalt, 78% anorthosite. Compare with Column (1).
(3) Hypothetical parent liquid for Apollo 11 igneous rocks derived from fractional crystallization model (Case 1 of Gast et al., 1970 renormalized).
(4) Hypothetical parent liquid for partial fusion model (Gast, 1972), renormalized.
(5) Deep interior (near liquidus crystals); Allende inclusions minus column (1) (this paper).
(6) Deep interior, partial melt model (Gast, 1972); assumptions (a) CaO and Al_2O_3 contents must be low and (b) lunar basalts are derived from great depth in a single stage process.

The composition of the outer layer of the Moon, obtained by removing the near liquidus crystals, is given in column (1) of Table V. For comparison, column (2) gives a previous 'average' crustal composition of the Moon. The similarity is remarkable. Column (3) gives the hypothetical parent liquid calculated by Hubbard et al. (1972) on the basis of trace element distributions and a fractional crystallization model. The MgO in column (3) is arbitrary because of lack of information regarding the extent to which olivine is involved in the source region. On the other hand the MgO content of the early condensates is also uncertain; it increases with falling temperature due to the increasing akermanite content of the melilite and rises rapidly once olivine and enstatite start condensing. The Moon may have accreted from material that condensed over a slightly broader temperature than the Allende inclusion. Column (4) is a hypothetical parent liquid derived from a partial melt model. The agreement with this model is not as good as the fractional crystallization model but the inferred Al_2O_3 and CaO content is still considerably greater than models such as Ringwood and Essene (1970). Column (5) is the inferred deep interior (>250 km) composition (spinel + melilite). For comparison column (6) gives Gast's (1970) deep interior composition which is based on the (invalid) constraint on total CaO and Al_2O_3. The density of the spinel-melilite assemblage is $3.2 \, \text{g cm}^{-3}$, about the same as the mean density of the Allende inclusions but $\sim 10\%$ greater than the density of the residual melt. At high pressure akermanite breaks down to merwinite plus diopside with a density of 3.29 g cm^{-3}; a similar reaction presumably occurs for gehlinite. The assemblage spinel + merwinite + diopside is probably stable through most of the bulk of the Moon. This assemblage has a density of $3.40 \, \text{g cm}^{-3}$.

8. Formation and Differentiation of the Moon

The derivation of the lunar surface rocks could proceed from our assumed composition for the moon in several ways. The following is one possible sequence:

(1) The Moon accreted from the material that condensed from a cooling solar nebula prior to the condensation of significant amounts of iron. The uncondensed material in the vicinity of the accreting Moon, including most of the iron and the volatiles were removed by an intense solar wind or were swept up by the more massive and more favorably disposed Earth.

(2) The whole Moon was enriched in Ca, Al, Ti, U, Th and the REE by approximately the ratio of the fraction of the material that condenses before iron relative to chondritic or solar non-volatile abundances. It was depleted in Fe, Na, Rb, K and the more volatile elements.

(3) The initial mineralogy of the Moon was primarily melilite (solid solution of akermanite and gehlinite), aluminous clinopyroxene, diopside, spinel, anorthite and perovskite.

(4) The rapid accretion, the high initial temperatures which are a consequence of the accretion-during-condensation hypothesis and the high U and Th abundances (10-16 times chondritic) lead to early and extensive, and perhaps complete, melting.

(5) The near liquidus phases, spinel and merwinite, settle to the interior. These crystals constitute approximately 65% of the mass of the Moon which corresponds to the volume below some 250 km. The REE, Ba and Sr are not necessarily strongly fractionated at this stage between crystals and melts. The melt, in fact, may be slightly depleted.

(6) The residual liquid is approximately 80% anorthosite and yields anorthosite, upon further cooling, which presumably formed the protocrust and the highlands, and then pyroxenes from which the basalts were derived either directly or by partial melting after solidification. An alternate scheme could involve the complete crystalliza-

tion of the outer shell followed by remelting and separation of the basalt liquid. The high U and Th content of the outer shell permits this possibility.

(7) The small initial FeO content of the Moon was strongly concentrated in the residual melts and, therefore, concentrated in the outer 250 km of the Moon.

(8) The basalts and anorthosites could be derived either by partial melting or by fractional crystallization, or both, of the outer 250 km of the Moon. The composition of this shell is similar to that inferred by Gast (1972).

Gast (1972) ruled out fractional crystallization on the basis that parent liquids with more than 80% anorthite were implausible. Anorthite contents of the order of 80% are, however, a natural feature of our model. A further difficulty pointed out by Gast with the fractional crystallization model is that the abundance of such elements as Ba, Sr and REE in the source region must be 15–20 times that of the average chondrite. These high abundances are also an intrinsic feature of our model.

The lunar igneous rocks could be either the result of a single stage extensive fractional crystallization or partial melting process, or could result from a multiple stage process involving both. The variation in the properties of the lunar basalts suggest that several processes may have operated. Some of the basalts may have crystallized from the residual melt after the higher temperature crystals were removed by sinking and floatation and others may have formed by partial melting at depth after crystallization of the liquid residuum.

The high temperature minerals in this model are gehlinite-akermanite solid solutions (at low pressure), spinel and perovskite. The thickness of the residual melt is ~ 230 km of which ~ 30 km is potentially basaltic. The fractionation of iron into the residual melt provides a high conductivity outer shell. The lattices of the high temperature minerals, in this model *can* accommodate the large REE and other ions although information is unavailable on the distribution coefficients. The fact that the REE ions substitute readily for Ca^{++} suggests that they may be retained by gehlinite and perovskite. If the initial concentration of the REE in the Allende inclusion is appropriate for the bulk Moon and if the anorthosite-basalt mix discussed previously is appropriate for the outer shell of the Moon it follows that the melt, the parent liquid of the lunar basalts and anorthosites, is slightly depleted in REE, relative to the bulk Moon, rather than enriched as is the case when olivine and orthopyroxenes are the high temperature phases.

9. Inhomogeneous Accretion of the Moon

It has been suggested several times that the Moon accreted inhomogeneously. However, the motivation has been to enrich the outer layer, in particular, the source regions of the lunar igneous rocks, in Ca, Al, U, Th, Ba and the REE. It has been considered unlikely that the whole Moon could exhibit these properties. However, we have shown that the early condensates in general, and the Allende inclusions in particular, provide the necessary characteristics of the source region and do not violate the inferred properties of the deeper interior. No primitive layering is required by the geochemical and geophysical data but chemical zonation as implied by the in-homogeneous accretion hypothesis is a distinct possibility.

The chemical zoning that has been proposed has the interior enriched in FeO, MgO, SiO_2 and the volatiles relative to the exterior, which is enriched in CaO, Al_2O_3, U, Th and the REE. This is contrary to expectations based on inhomogeneous accretion directly within a condensing solar nebulae. In this case, the CaO and Al_2O_3 would in-crease with depth and SiO_2, MgO and FeO would decrease with depth. The initial distribution of the refractory trace elements, such as Ba, U, Th and the REE depends on the phases in which they concentrate upon condensation. If they occur primarily in the gehlenite, perovskite, and spinel they can be expected to be concentrated initially in the interior of the Moon. If they are concentrated in the pyroxenes they can be ex-pected to be brought in with the upper layers. If the Moon partially or totally melts upon or after accretion they will be redistributed according to distribution coefficients between the melt and the liquidus phases – gehlenite and spinel.

The present day gross chemical layering in the Moon would be about the same whether it resulted from inhomogeneous accretion or homogeneous accretion followed by fractional crystallization or partial melting. In the inhomogeneous accretion model the phases in the deep interior would be merwinite, diopside spinel and, possibly, perovskite and corundum, the early condensing phases or reaction products. If the Moon were ever totally molten the interior would also be melilite, or merwinite. These are the near liquidus phases and are denser than the residual melt. In the partial melt model the low melting point and low density phases are pyroxenes and anorthite, which would rise to the surface to form the source region for the lunar basalts and anorthosites.

A critical test of the alternates involves the distribution of FeO. In the inhomogeneous accretion model the FeO would be concentrated near the surface because of the late condensation of iron. In the fractional crystallization model the residual melt, and hence the surface layers, would also be strongly enriched in FeO. In the homogeneous accretion, partial melt model the melt would be only slightly enriched in FeO. The main evidence bearing on this point, although controversial, is the conductivity profile of Sonett et al. (1971). They found a three orders of magnitude drop in electrical con-ductivity between 250 and 350 km depth, although other interpretations are possible. This can be interpreted in terms of a decrease in the FeO content at this depth. The mass fraction of the Moon below 250 km is 0.6 which is also the amount of gehlenite and spinel (the early condensates and also the near liquidus phases) in the Allende inclusions. Thus the inhomogeneous accretion and fractional crystallization models satisfactorily account for the gross layering. The outer 250 km in either of these models would be the source region from which the lunar igneous rocks are subsequently derived by partial melting or fractional crystallization. This source region must be enriched in U, Sr and the REE, relative to chondrites. If these are concentrated in the early condensates the deep interior must have been involved in the early and extensive differentiation, and this would favor fractional crystallization on the grand scale envisaged by Wood et al. (1970). If these trace refractories are concentrated in the later condensates, the interior need not be involved in a major way in the subsequent evolution of the Moon.

There are two possible variants of the direct heterogeneous accretion hypothesis. If accretion is rapid and completely efficient in the sense that it keeps up with the condensation, the moon will grow as a chemically zoned body with successive condensates shielding the early condensates from further reaction with the gas. One would obtain a Moon composed of a corundum nucleus overlain by perovskite, melilite and diopside shells. More likely some of the early condensate will be available for later reaction with the gas either before accretion or at the lunar surface. In this case, the Moon will be composed of diffuse shells grading from a primarily corundum, perovskite, melilite interior to a spinel rich shell overlain by diopside. The diopside and the spinel can react to form anorthite. The moon is unlikely to be perfectly prompt or efficient in accreting material that has condensed in its vicinity and it may, therefore, be initially a relatively homogeneous mixture of, primarily, perovskite, melilite, spinel and diopside, with, if temperatures fell low enough, some olivine. The amount of olivine is constrained to be small since its condensation interval overlaps iron and only a small fraction of the available (solar) iron has been incorporated into the Moon. For example, in a cooling gas of solar composition at 10^{-3} atm total pressure, 46% of the iron had condensed before forsterite appears (Grossman, 1972). More olivine may be incorporated into the interior if the condensation of iron is delayed by the non-equilibrium considerations of Blander and Katz (1967). As discussed previously, the melilite will break down to pyroxene plus merwinite at high pressure.

10. The Origin of the Moon

If the bulk of the Moon does represent a high temperature condensate the question arises, why did the Moon not accrete substantial quantities of material that condensed at lower temperatures? There are several possibilities.

The temperature at which an element or compound condenses out of a cooling nebula depends both on the composition and the pressure of the gas. The temperature of the nebula dies off rapidly away from the Sun and slowly with distance from the median plane. Pressure dies off with distance from the Sun and rapidly with height above the plane. At any given time the composition of the condensed material, prior to complete condensation, is a function of location in the nebula. If the uncondensed gas is removed at some stage, the planets and meteorites will differ in composition.

The difference in mass and composition of the Earth and the Moon can be explained if:

(a) The Earth and the Moon accreted from material that condensed at different distances from the median plane. Condensed material falls rapidly to the plane and forms the planetary nuclei which, initially, are refractory. In an isothermal nebula condensation occurs first in the median plane. As cooling continues more volatile material from near-plane and refractory material from off-plane is added. The Moon is assembled from the later off-plane refractory condensates that escape capture by the

Earth. In an adiabatic nebula condensation occurs first off-plane but planet assembly takes place at midplane. The Moon is formed from late condensing mid-plane refractories.

(b) The Moon was always in a low inclination orbit around the Sun and orbiting the Earth as well. Its encounter velocity with solar orbiting gains is higher than the Earth's and, therefore, it has a lower collection efficiency (Ganapathy *et al.*, 1970).

(c) The Earth started accreting sooner than the Moon, or for some other reason grew faster. It would therefore have a larger capture cross section than the Moon. When it became large enough to retain an atmosphere or, equivalently, to make a significant perturbation in the pressure of its surrounding gas envelope, it would retain infalling material more efficiently.

The observational fact that the Earth is bigger than the Moon and is enriched in iron and the volatiles compared to the Moon suggests that the Earth was more favorably disposed to collect the later condensates, and was possibly more favorably disposed throughout its accretional history.

11. Summary

The enrichment of refractories in the Moon such as Ca, Al, Ti, Ba, Sr, REE and U and the depletion of 'volatiles' such as Fe, Rb, K, S and H_2O relative to solar or carbonaceous chondritic abundances can be understood if the Moon represents a high temperature (pre-iron) condensate. The pre-iron condensates represent about 6% of the total condensables (exclusive of H, S and C) and will, therefore, be enriched in the refractory trace elements (such as Ba, Sr, REE, U and Th) by a factor of about 16, relative to carbonaceous chondrites. This is close to the average enrichment observed in the lunar surface material and in the Ca–Al rich inclusions of type II and III carbonaceous chondrites. The bulk surface chemistry of the Moon is consistent with the composition of the low-melting fraction of the early condensables. Trace elements, seismic and heat flow data are consistent with 'enrichment' of Ca, Al and U at the surface and in the interior of the moon. A Ca–Al rich deep interior does not imply an unacceptably large mean density. Most of the Moon's complement of volatiles may be brought in by chondritic material in the terminal stages of accretion.

The low melting fraction of the Ca–Al rich inclusions in the Allende meteorite provide a source region some 250 km thick which has the properties inferred by Gast (1972) to be appropriate for the source region of lunar basalts. The mantle, in this model, is primarily merwinite, diopside and spinel with a zero-pressure density of ~ 3.4 g cm^{-3}. The Moon in no way resembles a carbonaceous chondrite. One specific prediction of the model which can be tested by geophysical techniques is the gross layering, namely, a 250 km thick FeO rich crust overlying an iron poor mantle having a density of about 3.4 g cm^{-3}. The thermal and seismic aspects of this hypothesis are discussed by Anderson and Kovach (1972), Hanks and Anderson (1972), Anderson and Hanks (1972), and Anderson (1972).

The seismic profile is likely to be complex. The crust will be ~ 250 km thick but it may contain a layer, above ~ 90 km, which is in the garnet stability field and will therefore have higher density and velocity that the deeper and shallower parts of the crust (see Figure 8). Below 90 km the lunar temperature crosses the 'garnet in' curve; the lower part of the crust (90–250 km) should have relatively low seismic velocities. The velocities in the mantle should be high, around 8.5–9 km s^{-1}, as long as temperatures are below the solidus. Thermal history calculations (Hanks and Anderson, 1972) indicate that the deep interior may be above the solidus and, therefore, at least partially molten. This can occur anywhere below about 200 km and not be inconsistent with the shape of the Moon or the presence of mascons (Anderson and Hanks, 1972). The electrical conductivity profile suggests that partial melting, if it occurs, starts below 250 km. Seismic velocities decrease abruptly upon partial melting and attenuation, particularly for shear waves, increases.

12. Cameron's Objections

Cameron (1973) has objected to various cosmological implications of accreting the Moon and the Earth from material that condensed at different distances from the ecliptic. In particular he feels that the pressures inferred for the formation of the Earth (Figure 1) implies large densities at midplane and therefore considerable damping for objects in inclined orbits which must pass through this plane twice per orbital period. He also believes that, at midplane the objects will pick up iron.

First, the inferred pressure for the Earth's formation is not the ambient pressure at midplane. A massive body will perturb the gas pressure in its vicinity. For a body the size of the Earth the gas pressure at the surface will be about three orders of magnitude greater than ambient pressure. In addition, there are pressure fluctuations around the body due to aerodynamic forces. Finally, a large body will be outgassing trapped volatiles and have a transient increment to the atmosphere by each new impact.

The effective pressure and temperature in the vicinity of a large planet are greater than ambient. The presence of an atmosphere also makes a planet more efficient at scavenging and retaining already condensed material along its orbit. A late forming Moon whether in or out of the plane will be composed of only the earliest condensates. The hypothesis that the Moon or the Earth formed from material that condensed off the plane in no way requires that the body itself was fully assembled off the plane. Therefore, Cameron's objections are not relevant.

Forming the Moon from material that condensed at a later time has the advantage that it delays the start of the Moon, relative to the Earth, in a straight forward manner. It also explains, in a natural way, the inclination of the lunar orbit suggested by orbital evolution calculations. Capture from an Earth-orbit crossing trajectory is certainly more appealing than the double perturbation that is required if the Moon formed inside the orbit of Mercury.

Acknowledgement

This research was supported by National Aeronautics and Space Administration Contract NASA NGL 05-002-069.

References

Anderson, D. L.: 1972a, *Earth Planetary Sci. Letters*, in press.

Anderson, D. L.: 1972b, *Science*, in press.

Anderson, D. L.: 1972c, *Nature* **239**, 263.

Anderson, D. L.: 1972d, *J. Geophys. Res.*, in press.

Anderson, D. L. and Kovach, R. L.: 1972, *Phys. Earth Planetary Interiors*, in press.

Blander, M. and Katz, J.: 1967, *Geochim. Cosmochim. Acta* **31**, 1025.

Cameron, A. G. W.: 1973, *The Moon* **7**, 377.

Clark, S. P., Turekain, K. K., and Grossman, L.: 1972, in E. C. Robertson (ed.), *The Nature of the Solid Earth*, McGraw-Hill, 3–18.

Clarke, R., Jarosewich, E., Mason, B., Nelen, J., Gomez, M., and Hyde, J. R.: 1970, *Smithsonian Contrib. Earth Sci.* **5**.

Ganapathy, R., Keays, R., Laul, J. C., and Anders, E.: 1970, *Geochim. Cosmochim. Acta, Suppl. 1*, 1117.

Gast, P. W.: 1972, *The Moon* **5**, 121–128.

Gast, P., Hubbard, N., and Weismann, H.: 1970, *Geochim. Cosmochim. Acta, Suppl. 1*, 1143.

Green, T.: 1970, *Phys. Earth Planetary Interiors* **3**, 441.

Grossman, L.: 1972, *Condensation, Chondrites and Planets*, Ph.D. Thesis, Yale University, 97 pp.

Hanks, T. and Anderson, D. L., *Earth Planetary Interiors*, in press.

Hays, J.: 1966, *Carnegie Institution of Washington Year Book* **65**, 234.

Hoyle, F. and Wickramasinghe, N.: 1968, *Nature* **217**, 415.

Hubbard, N., Meyer, C., and Gast, P.: 1973 *Earth Planetary Interiors*, in press.

Hubbard, N., Gast, P., Meyer, C., Nyquist, L., Shih, C., and Weismann, H.: 1971, *Earth Planetary Sci. Letters* **13**, 71.

Ito, K. and Kennedy, G.: 1971, in J. Heacock (ed.), 'The Structure and Physical Properties of the Earth's Crust', *Am. Geophys. U., Geophys. Mono.* **14**, 303.

Kushiro, I.: 1964, *Carnegie Institution of Washington Year Book* **63**, 84.

Larimer, J.: 1967, *Geochim. Cosmochim. Acta* **31**, 1215.

Laul, J., Morgan, J., Ganapathy, R., and Anders, E.: 1971, *Proc. 2nd Lunar Sci. Conf.* **2**, 1159.

Lord III, H. C.: 1965, *Icarus* **4**, 279.

MacGregor, I.: 1970, *Phys. Earth Planetary Interiors* **3**, 372.

Marvin, U., Wood, J., and Dickey, J.: 1970, *Earth Planetary Sci. Letters* **7**, 346.

Mason, B. and Melson, W.: 1970, *The Lunar Rocks*, Wiley-Interscience, 179 pp.

Onuma, N., Clayton, R., and Mayeda, T.: 1972, *Geochim. Cosmochim. Acta* **36**, 169–188.

Prince, A.: 1954, *Am. Ceramic Soc. J.* **37**, 402.

Ringwood, A. E. and Essene, E.: 1970, *Science* **167**, 607.

Sonett, G., Shubert, G., Smith, B., Schwartz, K., and Colburn, D.: 1971, *Proc. Apollo 12 Lunar Sci. Conf.*, MIT Press.

Wakita, H. and Schmitt, R.: 1970a, *Nature* **227**, 478.

Wakita, H. and Schmitt, R.: 1970b, *Science* **170**, 969.

Wood, J., Dickey, J., Marvin, U., and Powell, B.: 1970, *Proc. Apollo 11 Lunar Sci. Conf.* **1**, 965.

THE GIANT PLANETS

W. B. HUBBARD

Dept. of Planetary Sciences, Lunar and Planetary Laboratory,
University of Arizona, Tucson, Ariz. 85721, U.S.A.

Abstract. The result that the giant planets are composed primarily of hydrogen and helium and lighter elements can be placed on a quantitative basis for Jupiter and Saturn because of the accuracy with which relevant equations of state are presently known. Current results are consistent with solar composition for Jupiter and perhaps also for Saturn. New theoretical techniques for calculating the gravitational multipole moments of rotating liquid planets may provide a way of applying future spacecraft data as strong constraints on interior structure.

1. Introduction

Placing the giant planets of the solar system in the context of the most important current astrophysical problems, it is clear that a reliable and precise result for their interior chemical composition is of utmost concern. As in the case of stars, to reach this goal requires obtaining an understanding of the interior structure and temperature distribution of these planets, a goal which is of course not without interest in itself since it bears upon possible modes of origin of these objects.

The low mean density of the giant planets, of the order of one $g\ cm^{-3}$, has for many years served to set them aside from the terrestrial planets, in a way just as fundamental as the differences in masses and rotation rates. It is clear that we are concerned, in the giant planets, with objects composed primarily of the lighter elements, and consequently with what may be, to some extent, undifferentiated samples of the primeval solar nebula. Intuition suggests, and detailed calculation confirms, that the properties of light elements such as hydrogen and helium may be somewhat more easily calculated in the dense liquid and solid state than the properties of the heavier, more electron-rich elements. As a result, the difficulty of calculating interior models of planets such as Jupiter and Saturn, and to a lesser extent, Uranus and Neptune, may be intermediate between modeling the Sun, a gaseous object, and the terrestrial planets. This provides some compensation for the difficulty of obtaining *in situ* observations of the giant planets due to their remoteness. Considerable theoretical, experimental and observational progress has been made in the understanding of Jupiter and Saturn, in particular, during the past decade. In this paper, we will discuss some of the most important of these advances, as well as some recent developments which could be exploited by observations from spacecraft.

A major point which we shall develop is that accurate determination of the gravitational multipole moments out to terms of very high order may be a fundamental key to determination of the interior density distribution and hence to quantitative determination of the bulk chemical composition.

A. G. W. Cameron (ed.), Cosmochemistry, 153–163. All Rights Reserved

2. Observational Constraints on Interior Chemical Composition

The most nearly model-independent result which one can derive for the bulk composition of a planet is through comparison of the mean density with that of a homogeneous, one-component sphere in hydrostatic equilibrium at zero temperature. In such a calculation, the only theoretically uncertain quantity is the zero-temperature equation of state. The most recent paper on this subject was published by Zapolsky and Salpeter (1969), using Thomas-Fermi-Dirac equations of state. Figure 1 shows the results of Zapolsky and Salpeter's calculations, with observed points for the major planets

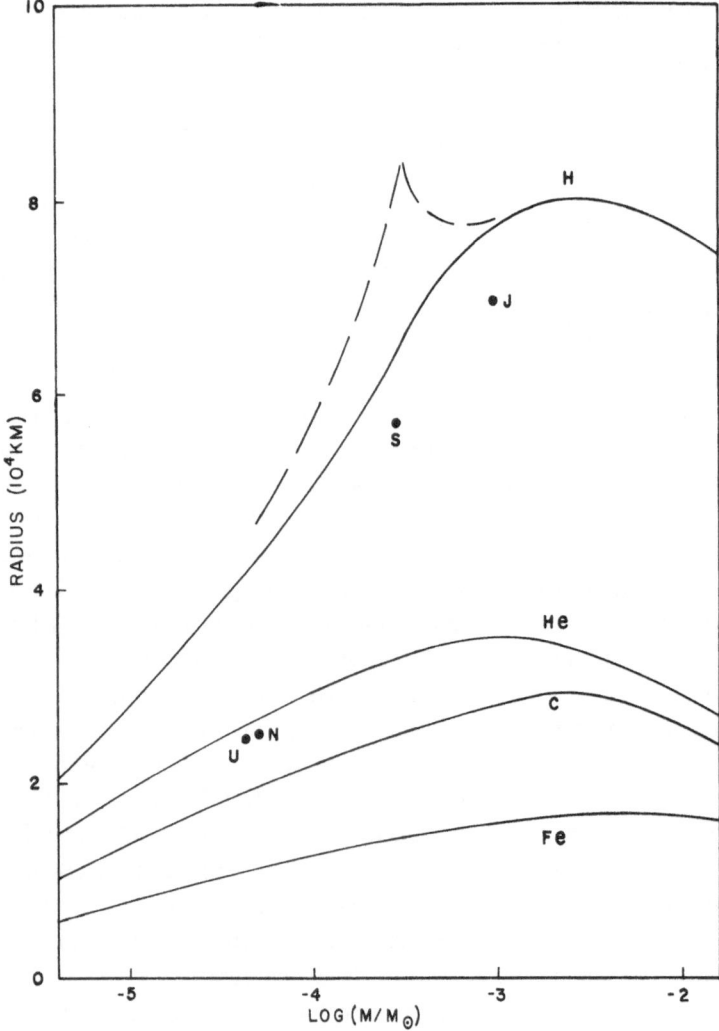

Fig. 1.   Theoretical mass-radius diagram with observed points for the giant planets (J, S, U, N), *Solid curves*, from the Thomas-Fermi-Dirac theory of Zapolsky and Salpeter (1969). *Dashed curve*, approximate correction to hydrogen curve from inclusion of molecular hydrogen envelopes.

plotted. The conclusion which one draws from this diagram is identical to that drawn from earlier studies (Abrikosov, 1964; DeMarcus, 1958), namely, that Jupiter and Saturn must have large fractions of hydrogen and helium. Unfortunately, the situation for heavier elements is not so clear since the curves for these elements are not as well separated as for hydrogen and helium. As a result, no firm conclusions can be reached about the internal composition of Uranus and Neptune, although a literal interpretation of their position on Zapolsky and Salpeter's diagram suggests that they are composed of helium or carbon.

In order to make quantitative the above qualitative conclusions, it is necessary to impose as many observational constraints as possible on interior models, and to include effects such as rotation and finite interior temperature. (As is apparent from Figure 1, only in the case of Jupiter and Saturn is there reasonable hope at present of obtaining an unambiguous answer for the interior composition. Therefore, we will concentrate in this paper mainly upon these objects.)

Jupiter and Saturn are the most rotationally flattened objects in the solar system, with oblatenesses of 6% and 10%, respectively. This effect must be included in planetary structure calculations. Specifically, if we know the observationally accessible parameter

$$\alpha = R^3 \omega^2 / GM,\tag{1}$$

where R is the equatorial radius of the planet, ω is its angular rotation velocity, G is the gravitational constant, and M is the mass of the planet, then the gravitational moments $J_2, J_4, J_6, \ldots$ are functions only of α and the internal mass distribution. For Jupiter and Saturn, α, J_2 and J_4 are available (Peebles, 1964), giving constraints on interior models.

Another observational parameter of interest is the atmospheric chemical composition. Certainly for the terrestrial planets, the atmospheric composition is at best only distantly related to the bulk composition of the planet. For Jupiter, on the other hand, there exists evidence that the atmosphere is predominantly hydrogen (Owen, 1970; Hubbard et al., 1972), suggesting that the interior may be similar in composition to the atmosphere. Observational data are much poorer for Saturn's atmosphere, but it is probable that it is similar in composition to Saturn's interior.

It turns out that the thermal expansion coefficient for high-pressure hydrogen and helium mixtures is rather high. Consequently, interior temperatures comparable to terrestrial interior temperatures ($\sim 5000\,\mathrm{K}$) would produce considerable thermal perturbations to interior density distributions in Jupiter and Saturn, and thus have a serious effect on the deduction of interior chemical composition. The best observational constraint on internal temperatures for Jupiter and Saturn is provided by the infrared data of Aumann et al. (1969), who have derived a net heat flow of 13 000 erg cm^{-2} s^{-1} and 3000 erg cm^{-2} s^{-1} for Jupiter and Saturn, respectively. No heat flow data are available for Uranus and Neptune, which greatly limits progress in understanding their interior structure.

3. Input Physics and Models

Considerable progress has been made recently on high-pressure equations of state, particularly in the case of hydrogen. In particular, it has become evident that a variety of theoretical approaches give substantially equivalent results for the equation of state of metallic hydrogen, the pressure-ionized form of solid hydrogen.

Figure 2 shows a theoretical phase diagram for hydrogen. Under conditions which obtain in Jupiter and Saturn, temperatures are always well below the degeneracy temperature of the electrons ($\sim 500\,000$ K); consequently, the electrons are always in

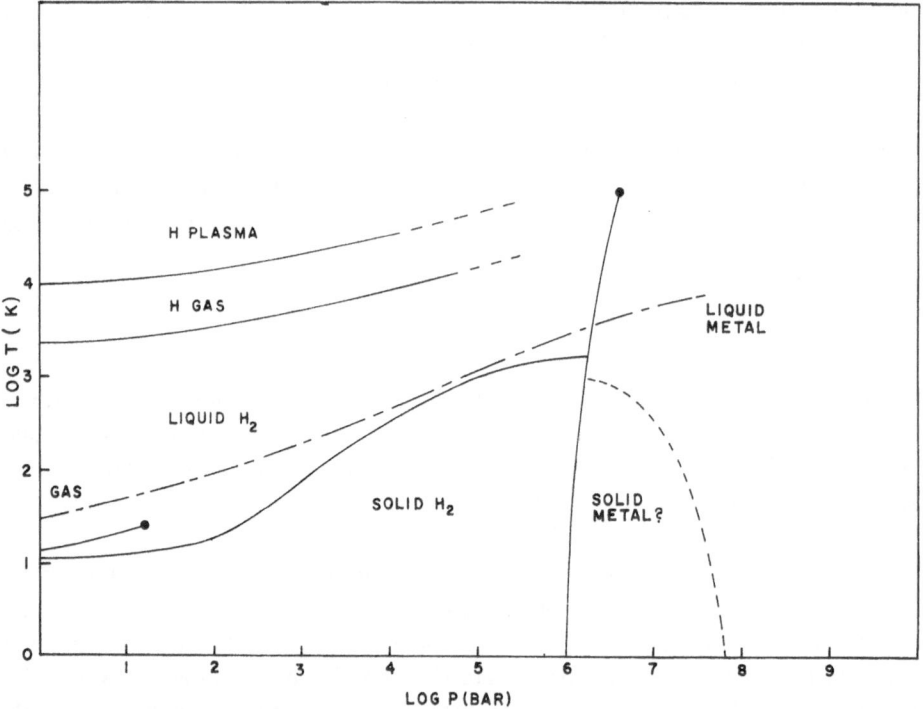

Fig. 2. Theoretical phase diagram for hydrogen. The melting curve for solid H_2 is based upon recent calculations by Slattery and Hubbard (1972) and is uncertain by about 20%. Solid metallic hydrogen forms only at rather low temperatures and may be entirely unstable due to large quantum zero-point vibrations. *Adiabat* is appropriate for the Jovian model of Hubbard (1970).

their ground state, which simplifies the calculation somewhat. The problem is to find the ground state electron wave function in the potential of the protons. The key to solving this problem is to derive solutions in the high- and low-pressure limits and then bridge the gap using experimental data.

In the limit of high pressure, the electron Fermi energy dominates the energy of the system and the electrons are essentially uniform. Provided that the pressure is not too high, the protons are in a lattice and the energy and pressure can be calculated with high accuracy ($\sim 1\%$) (Salpeter, 1961).

At densities which obtain in Jupiter and Saturn (~ 1 g cm^{-3}), the electron Fermi energy is comparable to the energy of interaction between electrons and protons, and the deviation of electron density from uniformity must be taken into account. There are three theoretical techniques for handling the equation of state in this region:

(a) *Wigner-Seitz Method* (DeMarcus, 1958)

The protons are assumed to be in a lattice. A single lattice site is assumed to be replaced with a spherical cell with the proton at the center. The electron density is calculated from a solution to the Schrödinger equation in the spherical cell, subject to continuity at the surface of the cell.

(b) *Thomas-Fermi-Dirac Method* (Salpeter and Zapolsky, 1967)

Proceeds as in Wigner-Seitz theory except that the electron density is calculated by requiring the electron chemical potential to be constant throughout the cell.

(c) *Dielectric Function Theory* (Hubbard and Slattery, 1971)

The protons can be in an arbitrary configuration. The electron density is calculated in a self-consistent manner from the linear response of the electrons to the proton Coulomb field.

The three techniques give results for the energy and pressure of solid metallic hydrogen which agree to within 5% for densities greater than 1 g cm^{-3} (Hubbard, 1972a). As a result, the equation of state of solid metallic hydrogen is very probably not a serious source of uncertainty in model planets.

A great advantage of the dielectric function approach is that it can be applied to arbitrary ion configurations and thus is a tractable procedure for investigating the liquid phase of metallic hydrogen. Using methods developed by Metropolis *et al.* (1953), Hubbard and Slattery (1971) have calculated thermodynamic variables for metallic hydrogen. Their results indicate that metallic hydrogen under Jovian conditions is liquid at temperatures in excess of ~ 3000 K.

The calculation of thermodynamic quantities at densities below 1 g cm^{-3} (pressures below ~ 2 Mbar) becomes uncertain, and techniques begin to diverge. Instead, we rely on experimental data in this region. Let us suppose that a pair potential for molecular hydrogen is available. For example, calculations by Tapia *et al.* (1971) have been performed for the interaction energy of two isolated hydrogen molecules. Such calculations can be supplemented by experimental virial coefficient data (Friedman, 1957), to help determine the location and depth of the potential energy minimum. The results of such a procedure indicate that a potential of the exp -6 variety gives a good fit to low-pressure data (Slattery and Hubbard, 1972) and also to high-pressure data (Bender *et al.*, 1972). A very stiff potential of the Lennard-Jones 6-12 variety is definitely inconsistent with presently available data.

The approach which is suggested, then, is that we use the H_2-H_2 intermolecular potential to calculate the energy of molecular hydrogen up to multimegabar pressures, where the tightly bound electron states abruptly go over to a weakly bound metallic phase in a first-order phase transition. Such a transition can be calculated using the accurate metallic equation of state and the less-accurate molecular equation of state. The current value for the transition pressure at zero temperature is in the range 1–5

Mbar (Trubitsyn, 1971). An experimental determination of the transition pressure seems feasible (Hawke *et al.*, 1971), but has not yet been carried out. Since the transition pressure is highly sensitive to small changes in the molecular equation of state, it is likely that even a low accuracy measurement of the transition pressure ($\pm 25\%$) would pin down the molecular equation of state very precisely.

To summarize the situation for hydrogen, the metallic hydrogen equation of state is in very good shape. The molecular hydrogen equation of state is poorer, but it is considerably better determined now than it was ten years ago (see, e.g., DeMarcus, 1958).

The problem of calculating the equation of state of a mixture of hydrogen and other elements is a critical one for determining the chemical composition of the giant planets. The method which is usually adopted (DeMarcus, 1958; Peebles, 1964; Hubbard, 1969, 1970) can be called the additive volume approximation, and is a simple analytic procedure for combining the zero-temperature equations of state of pure subcomponents. Let ϱ be the density in g cm^{-3} of the chemical mixture, and let $\varrho_i(P)$ be the density that the pure ith subcomponent would have a pressure P. The equation of state $\varrho(P)$ of the mixture is then given, in the additive volume approximation, by

$$\varrho - 1 = \sum_i X_i \varrho_i^{-1}(P), \tag{2}$$

where X_i is the mass fraction of the ith subcomponent.

Whereas Equation (2) is essentially correct for inmiscible subcomponents, its general validity is doubtful, and its generalization to finite temperature is obscure. Nevertheless, it is important to realize that all of the information which is presently available about the H/He ratio in Jupiter and Saturn, from interior models, presumes the validity of Equation (2).

It is possible to test Equation (2) using dielectric function theory. In this approach, one performs ensemble averages on a system of protons and alpha-particles with a background uniform electron distribution, and electron screening included to first order.

Under Jovian conditions, this procedure is accurate for the protons but only marginally valid for the alpha-particles because of the much stronger electron-ion interaction. Calculations (Hubbard, 1972a) indicate that as zero temperature is approached in a hydrogen-helium plasma, the additive volume law is approached with high accuracy, for pressures greater than one Mbar. The physical picture seems to be that the protons and alpha-particles establish 'spheres of influence' about themselves in accordance with this law as low temperatures are reached. Table I gives some results of the hydrogen-helium calculations.

It should be noted that the additive volume law would be satisfied in the event of inmiscibility of the hydrogen-helium system. No convincing evidence of inmiscibility at temperatures of ~ 5000 K and higher exists so far, although there are hints of a phase separation at lower temperatures, at 12 Mbar (Hubbard, 1972b).

It must be remembered that the calculations are done in a unit cell with a small

TABLE I

Equation of state for hydrogen=helium mixtures

$\rho(\text{g cm}^{-3})$	$T(\text{K})$	$P(\text{Mbar})$	γ
(Solar Composition: 73% H, 27% He by mass)			
1.09	20900	2.63	0.66
1.09	5220	1.20	0.67
1.09	2780	0.93	0.68
2.13	26100	12.13	0.65
2.13	6520	8.53	0.65
2.13	3480	7.88	0.64
3.68	31300	36.71	0.64
3.68	7820	29.32	0.64
3.68	4170	27.95	0.64
(50% H, 50% He by mass)			
3.10	26100	16.90	0.65
3.10	6520	12.62	0.64
3.10	3480	11.84	0.64
5.36	31300	52.12	0.64
5.36	7820	43.36	0.62
5.36	4170	41.72	0.62

number (~ 40) of particles, and the periodic boundary conditions prevent any sizeable phase separation. More elaborate calculations are presently underway, with more particles, to examine this possibility. Needless to say, if hydrogen and helium are in fact inmiscible in the interior of Jupiter and Saturn, there will be a profound effect on the interior structure and on the atmospheric H/He ratio (Smoluchowski, 1967).

Modeling of Jupiter and Saturn has not kept up with the finite-temperature equations of state described above. When results become available for mixtures of molecular hydrogen and helium (Hubbard and Slattery, in progress), a new round of model building can begin. Earlier models of Jupiter and Saturn (Hubbard, 1969, 1970) were based upon the result (Hubbard, 1968) that the heat flux observed by Aumann *et al.* (1969) implies a liquid, convective metallic hydrogen interior. This result, which is consistent with the more recent thermodynamic results, implies adiabatic structure in the metallic portion of the planet, and also gives a reasonable explanation of the magnetic field of Jupiter. If Saturn has a metallic hydrogen core, it should also have a magnetic field.

If we assume that the molecular envelopes of Jupiter and Saturn are also adiabatic and convective, which is not in contradiction to any available observational data (Gulkis and Poynter, 1972), the thermal structure of the planet is readily calculated from the Gruneisen parameter:

$$\gamma = (\partial \ln T / \partial \ln \varrho)_s, \tag{3}$$

where T is the temperature, and the derivative is taken along an adiabat. Various estimates of γ for a hydrogen plasma under Jovian conditions have been made (Hubbard, 1969, 1970; Trubitsyn, 1971; Kopyshev, 1965), leading to values in the range 0.5–0.6. The Monte Carlo calculations of Hubbard and Slattery (1971) indicate $\gamma = 0.64$, and this value is not greatly affected by a small admixture of helium (Hubbard, 1972a). In the molecular phase, γ has a similar value (Slattery and Hubbard, 1972).

In adiabatic models, the central temperature is thus defined by the upper terminus of the adiabat, which is observationally accessible. The most recent value of the central temperature of Jupiter and Saturn, approximately the same for both planets, is 7000–8000 K (Hubbard, 1969, 1970). This leads to large thermal perturbations to the central pressure, $\sim 10\%$ for Jupiter, and $\sim 20\%$ for Saturn, and consequently increases the abundance of helium over zero-temperature models. It should be pointed out that Hubbard's models do not include constituents heavier than helium. It is conceivable that such elements could be overabundant relative to a solar mix, with normal H/He, and still produce an acceptable model (Cameron, 1972).

With the above provisos, Jupiter is found to consist of $\sim 65\%$ hydrogen by mass, with the rest mostly helium. The error bars on such a determination are mainly due to the uncertainty of the central temperature and the molecular hydrogen equation of state. A reasonable estimate for the error is $\pm 10\%$, which would make the result consistent with solar composition. A literal extension of the same procedures to Saturn gives the result that this planet should be mostly helium by mass despite its lower mean density. However, Saturn should have only a rather small core of metallic hydrogen, and uncertainties in the equation of state are consequently severe.

4. Gravitational Multipole Moments

As we have seen, there are reasons for believing that Saturn and Jupiter are entirely fluid objects. Consequently, we may expect potential theory and the assumption of hydrostatic equilibrium to provide significant constraints on the internal structure of these objects, and thus on their internal composition.

The external gravity potential of a planet in hydrostatic equilibrium is given by

$$\varphi(r, \mu) = (- GM/r) \left[1 - \sum_{l=1}^{\infty} J_{2l}(R/r)^{2l} P_{2l}(\mu) \right], \tag{4}$$

where G is the gravitational constant, M is the mass, r is the distance from the center of the planet, μ is the cosine of the angle from the rotation axis, R is the equatorial radius of the planet, P_{2l} are the Legendre polynomials, and J_{2l} are the multipole moments which depend on α (Equation (1)) and the mass distribution in the planet.

Presently only J_2 is known with any precision for Jupiter and J_2 and J_4 for Saturn. Values of J_2 are also available for Uranus and Neptune. It is likely that spacecraft orbiters and/or radar data from natural satellites will provide, later in this century, gravity data to much higher order.

The question thus arises of whether it will be possible to exploit such data in terms of better constraints on interior models.

Theories valid to order α^2, i.e., to order J_4, have been developed by DeMarcus (1958), and Peebles (1964). A third order theory, valid to order J_6 has been presented by Zharkov et al. (1971). Naturally, analytic procedures become very tedious at higher order, and no calculations have been attempted for J_8.

The theory of Zharkov et al. (1971) has been applied to models of Jupiter and Saturn by Zharkov et al. (1972), but only in a restricted sense, using as a starting point a given distribution of density with mean radius, $\varrho(s)$, rather than an equation of state $P(\varrho)$.

We will give here a preliminary report on an alternate numerical procedure for calculation of gravitational multipole moments to high order, using a scheme derived by Ostriker and Mark (1968). We consider a sphere which just contains the rotating planet, i.e., with radius equal to the equatorial radius. Within the sphere, we will seek to represent the mass density ϱ by an approximate polynomial expression of the form

$$\varrho(\chi, \mu) \simeq \bar{\varrho} \sum_{l=1}^{N} \sum_{m=1}^{N} A_{lm} \chi^{2m-2} P_{2l-2}(\mu),\tag{5}$$

where χ is the distance from the center of the planet in units of the equatorial radius R, μ is the cosine of the angle to the polar axis, $\bar{\varrho}$ is the mean density within the sphere $3M/4\pi R^3$, P_l are the Legendre polynomials, and A_{lm} are dimensionless coefficients. Here N is the order of the expansion which can be chosen arbitrarily to give a satisfactory fit to the density distribution.

The method of Ostriker and Mark thus represents the density in terms of a finite matrix A_{lm} which can be calculated by matrix manipulation techniques to give a self-consistent gravity field for a rotating planet. The method works for any axisymmetric rotation field, and can thus be used to investigate the effect of differential rotation.

In terms of Equation (5), it is straightforward to show that the mass multipole moments are given by

$$J_{2l-2} \simeq - \sum_{m=1}^{N} 3A_{lm}/(4l-3)(2m+2l-1).\tag{6}$$

We note the following constraints on A_{lm}: A_{lm} with $m<l$ must be zero in order to satisfy continuity at the origin. Also, $\varrho(1, \mu)$ must be zero, which means that the sum of rows of the matrix must be zero. If we know the multipole moments of the planet, J_{2l}, this gives another constraint on a weighted sum of rows of the matrix. Clearly, this is not sufficient to uniquely define the matrix, but it does give a useful constraint on planetary interior models.

We have performed a number of calculations on Jovian models and polytropes with the self-consistent-field (SCF) program of Ostriker and Mark, which was modified for our purpose from a version kindly supplied by P. Bodenheimer. It appears that one can reliably calculate J_6 and possibly J_8 for Jupiter and Saturn with $N=15$, the maximum number of terms achievable without extensive reprogramming. It is important to remember however, that one could calculate to higher order with this method, should observational data justify the effort.

Initially, we considered planets rotating as a solid body. The effect of differential rotation was then put in through a rotation law of the form

$$\omega = \omega_c (1 + ca), \tag{7}$$

where ω_c is the central rotation frequency, a is the distance from the polar axis in units of R, and c is a dimensionless differential rotation parameter. Other differential rotation laws could also be used.

Table II presents results for a Jovian model calculated with the SCF program using the same equation of state $P(\varrho)$ as in model $J9$ of Hubbard (1970). All models have

TABLE II

Parameters of Jovian models

c	α	R(km)	J_2	J_4	J_6	N
0.00	0.091 605	71 794	0.014 367	−0.000 443	0.000 049	15
0.01	0.091 593	71 815	0.014 461	−0.000 450	0.000 050	15
0.01	0.091 620	71 811	0.014 443	−0.000 424	0.000 067	13
(observed)		71 880	0.014 71	−0.000 674	?	
		± 30	± 0.000 14	± 0.000 376		
(Zharkov et al.)		71 300	0.015 67	−0.000 661	0.000 041	

the same total angular momentum, and mean rotation periods are the same to within 50 s. For comparison, we also give observational data and results of the theory of Zharkov et al. (1972) applied to Hubbard's model $J7$ (1969) which is slightly more helium-rich than model $J9$.

We conclude:

(a) The error in J_6 from using a 15-term polynomial density expansion is of the order of 25%, as judged from the results of an $N=13$ expansion. Thus, more terms are probably necessary to calculate J_6 with higher accuracy for a given Jupiter model.

(b) The effect of 1% differential rotation is very small and can probably be ignored in theories of Jovian structure.

Greater accuracy should result for Saturn models since J_4 is larger with respect to J_2 in this case. Saturn may have as much as 11% differential rotation (Newburn and Gulkis, 1972), and it will be interesting to see the effect of this on gravitational moments in future calculations.

5. Conclusion

The advent of high-speed computers has permitted numerical experiments which give considerable insight into the behaviour of material under conditions typical of the interior of Jupiter and Saturn. Advanced experimental techniques also have provided significant advances. It is probably correct to say the we have entered the quantitative era in modeling the giant planets.

Some of the research reported in this paper was supported by NSF Grant GP-19667.

References

Abrikosov, A. A.: 1964, *Sov. Phys. – JETP* **18**, 1399.

Aumann, H. H., Gillespie, C. M., Jr., and Low, F. J.: 1969, *Astrophys. J.* **157**, L69.

Bender, C. F., Hoover, W. G., Olness, R. J., Rogers, F. J., and Ross, M.: 1972, Proc. Conf. on High-Pressure Phys. and Planetary Interiors, Houston, Texas, March 1–3, 1972, American Elsevier, New York.

Cameron, A. G. W.: 1972, private communication.

DeMarcus, W. C.: 1958, *Astron. J.* **63**, 2.

Friedman, A. S.: 1957, in A. A. Bennett *et al.* (eds.), *AIP Handbook*, McGraw-Hill, New York, pp. 4–158.

Gulkis, S. and Poynter, R.: 1972, Proc. Conf. on High-Pressure Phys. and Planetary Interiors, Houston, Texas, March 1–3, 1972, American Elsevier, New York.

Hawke, R. S., Duerre, D. E., Huebel, J. G., Keeler, R. N., and Klapper, H.: 1971, *Nature* **233**, 79.

Hubbard, W. B.: 1968, *Astrophys. J.* **152**, 745.

Hubbard, W. B.: 1969, *Astrophys. J.* **155**, 333.

Hubbard, W. B.: 1970, *Astrophys. J.* **162**, 687.

Hubbard, W. B.: 1972a, *Astrophys. J.* **176**, 525.

Hubbard, W. B.: 1972b, Proc. Conf. on High-Pressure Phys. and Planetary Interiors, Houston, Texas, March 1–3, 1972, American Elsevier, New York.

Hubbard, W. B. and Slattery, W. L.: 1971, *Astrophys. J.* **168**, 131.

Hubbard, W. B., Nather, R. E., Evans, D. S., Tull, R. G., Wells, D. C., Van Citters, G. W., Warner, B., and VandenBout, P.: 1972, *Astron. J.* **77**, 41.

Kopyshev, V. P.: 1965, *Sov. Phys. – Doklady* **10**, 338.

Metropolis, N., Rosenbluth, A. W., Rosenbluth, M. N., Teller, A. H., and Teller, E.: 1953, *J. Chem. Phys.* **21**, 1087.

Newburn, R. L., Jr. and Gulkis, S.: 1972, *Space Sci. Rev.* **3**, 179.

Ostriker, J. P. and Mark, W.-K.: 1968, *Astrophys. J.* **151**, 1075.

Owen, T.: 1970, *Science* **167**, 1675.

Peebles, P. J. E.: 1964, *Astrophys. J.* **140**, 328.

Salpeter, E. E.: 1961, *Astrophys. J.* **134**, 669.

Salpeter, E. E. and Zapolsky, H. S.: 1967, *Phys. Rev.* **158**, 876.

Slattery, W. L. and Hubbard, W. B.: 1972, *Astrophys. J.* **181**, 1031.

Smoluchowski, R.: 1967, *Nature* **215**, 691.

Tapia, O., Bessis, G., and Bratoz, S.: 1971, *Int. J. Quant. Chem.* **4**, 289.

Trubitsyn, V. P.: 1971, *Soviet Astron. – AJ* **15**, 303.

Zapolsky, H. S. and Salpeter, E. E.: 1969, *Astrophys. J.* **158**, 809.

Zharkov, V. N., Trubitsyn, V. P., and Makalkin, A. B.: 1972, *Astrophys. Letters* **10**, 159.

Zharkov, V. N., Trubitsyn, V. P., and Samsonenko, L. V.: 1971, *Fizika Zemli i Planyet*, Nauka Press, Moscow.

ABSTRACTS OF CONTRIBUTED PAPERS

METAL ABUNDANCE IN POPULATION I STARS

R. C. HENRY, J. E. HESSER, and W. McCLINTOCK

Dept. of Physics, Johns Hopkins University, Baltimore, Md. 21218, U.S.A.

The strength of the K line of singly ionized calcium has been measured for several hundred A-type stars within a few hundred parsecs of the Sun and for the A stars in several galactic star clusters. The derived abundance of calcium varies from star to star by up to a factor of 2, and there is no correlation of abundance with the space motion of the stars.

We have obtained new observations on this photometric system of the unreddened Coma star cluster, and these data will be presented and discussed.

NUCLEOCHRONOMETERS FOR THE p- AND s-PROCESSES*

J. AUDOUZE** and D. N. SCHRAMM†

California Institute of Technology, Pasadena, Calif. 91109, U.S.A.

Previous nucleochronometers have been nuclei of r-process origin. This present work shows that the pure p-process radioisotope ^{146}Sm ($\tau_{1/2} = 7 > 10^7$ yr) can be used as a chronometer for p-process nucleosynthesis. The estimated p-process production ratio of this isotope relative to ^{144}Sm is shown to be $0.35 \leqslant {}^{146}$Sm$/^{144}$Sm $\leqslant 0.6$. Since ^{146}Sm alpha decays to ^{142}Nd, there should be a variation in the isotopic abundance of ^{142}Nd in meteorites, depending on the Sm/Nd elemental ratio in some particular meteoritic mineral phases. It is estimated that ^{142}Nd isotopic anomalies of at least 10^{-3} can be found. Such a measurement would give the time interval between the last p-process nucleosynthetic event and the formation of solid bodies in the solar system.

A possible s-process chronometer is ^{176}Lu ($\tau_{1.2} \simeq 3 \cdot 10^{10}$ yr). However, because of its long lifetime and low abundance, it is impossible to determine an accurate chronology for this nucleus by use of currently available experimental techniques. For this case, the chronology argument can be reversed in that all reasonable nucleochronologies imply a narrow range in values for the neutron-capture cross sections and s-process abundances of the nearby s-process nuclei. The important branching ratio $\sigma[^{175}$Lu$(n, \gamma) {}^{176}$Lu$]/\sigma[^{175}$Lu$(n, \gamma)(^{176}$Lu $+ {}^{176m}$Lu$)]$ is predicted to be 0.33 ± 0.17.

CHRONOLOGY OF GALACTIC HEAVY-ELEMENT NUCLEOSYNTHESIS

T. P. KOHMAN and J. M. HUEY

Dept. of Chemistry, Carnegie-Mellon University, Pittsburgh, Pa. 15213, U.S.A.

Several models have been developed for the average rates of star formation, heavy-element nucleosynthesis, and destruction of heavy elements in the Galaxy, considered as a homogeneous medium initially consisting of hydrogen and helium. The models predict the time dependence of the relative masses of stars and interstellar matter and the heavy-element content of the latter. They also predict the abundance ratios of any two nuclides whose relative production rates are known, with particular reference to the solar system.

The influence of various model parameters on various observable quantities has been investigated. Evaluation of models and parameters is attempted by consideration of the nuclide ratios ^{235}U$/^{238}$U, ^{232}Th$/^{238}$U, ^{244}Pu$/^{238}$U, and ^{129}I$/^{127}$I in meteorites, the atmospheric composition of stars in dated globular and galactic clusters, and the present amount of interstellar gas and dust.

The observations are compatible with the model of continuous nucleosynthesis [1] decaying exponentially [2] modified by a terminal solar system spike and heavy-element destruction [3]. There

* Supported in part by the National Science Foundation grants GP-28027 and GP-27304.
** Supported in part by French-U.S. exchange fellowship.
† Present address: University of Texas, Austin, Texas.

is no need to invoke an initial spike [4] associated with the collapse of the Galaxy [5], or heavy-element dilution [6] by an influx of extragalactic gas [7], although neither is excluded by present data.

^{224}Pu AS A POSSIBLE INDICATOR OF INTERSTELLAR DUST WITHIN THE SOLAR SYSTEM*

G. A. COWAN

*University of California and Los Alamos Scientific Laboratory
Los Alamos, N.M. 87544, U.S.A.*

We propose to examine the extent to which ^{244}Pu (8.28×10^7 yr half-life) may enter the solar system and accrete in 'contemporary' materials. We hypothesize that a major fraction of the ^{244}Pu is condensed on interstellar grains and that a significant fraction of the grain mass occurs in sizes sufficiently large to reach Earth orbit ($> 0.3 - \mu$ radius). Uncertainty as to the size distribution of grains is probably the weakest link in this hypothetical chain.

Since the currently accepted value for ^{244}Pu/^{238}U in meteorites at condensation time [1] is compatible with the existence at the time of solar-system formation of secular equilibrium between heavy elements in the interstellar medium and a 'steady' supernova source, it is assumed that such secular equilibrium continues to exist. On this basis, upper limit fluxes are calculated at Earth orbit for heavy elements in interstellar grains. In addition, published cosmic-ray heavy-element data are used to estimate the cosmic-ray flux of ^{244}Pu. [2, 3, 4]

We find that between passages of the solar system through clouds the flux values of ^{244}Pu in cosmic rays and on grains may be comparable in size ($\sim 10^{-11}/$cm^{-2} s^{-1} at the top of the atmosphere), but the grain flux may be 10^4 times higher during a cloud passage. Evidence for such fluxes, particularly at the higher levels, should be preserved in certain terrestrial sedimentary deposits, e.g., phosphate rocks, and a time-averaged exposure level may be measurable on the Moon. The kinetic energy of the grains is no more than a few electron volts per nucleon, so they will adhere to but will not penetrate the lunar surface. A properly selected sample of lunar fines, possibly as small as 10 g, should provide a measurable quantity of ^{244}Pu if interstellar grains penetrate the solar system.

Plans are currently being made to search for ^{244}Pu in sedimentary phosphate rock and to amplify existing efforts to find ^{244}Pu in lunar material, with emphasis on careful collection and selection of surface samples.

References

1. Podosek, F. A. and Huneke, J. C.: *Earth Planetary Sci. Letters* **12**, 72 (1971).
2. Price, P. B., Fowler, P. H., Kidd, J. M., Kobetich, E. J., Fleischer, R. L., and Nichols, G. E.: *Phys. Rev. D.* **3**, 815 (1971).
3. Bell, G. I.: private communication.
4. Hoffman, D. C., Lawrence, F. O., Mewherter, J. L., and Rourke, F. M.: *Nature* **234**, 132 (1971).

NONEQUILIBRIUM CHEMISTRY AND THE COMPOSITION OF COSMIC CLOUDS

B. D. DONN, E. W. CHAPPELLE, W. A. PAYNE, and L. J. STIEF

*Astrochemistry Branch, Laboratory for Extraterrestrial Physics,
NASA/Goddard Space Flight Center, Greenbelt, Md. 20771, U.S.A.*

The molecular composition of cosmic clouds is nearly always determined by use of molecular equilibrium calculations. In this paper, we examine the validity of this procedure as applied to high-temperature regions – e.g., collapsing clouds and prestellar nebulae; expanding, cool stellar atmosphere, circumstellar shells; and colliding clouds.

---

* Work performed under the auspices of the U.S. Atomic Energy Commission.

Several factors affect equilibrium calculations: (1) rate of approach to equilibrium compared to the appropriate astronomical time scale, (2) spatially or temporally varying temperature of the system, (3) freezing out of reactions with high activation energies, (4) necessity of assuming final composition for which abundances are calculated, and (5) possible disequilibrium among molecular energy modes and its effect on reaction rates.

Because of these problems, we are carrying out experiments to simulate astrophysical systems. These experiments also provide basic data for estimating rates of approach to equilibrium and the actual composition. The extent of reaction of a $1:100$ $HCN:H_2$ mixture at 1100 K was determined at 2, 9, 48, and 100 hr. A lifetime (e^{-1}) of about 18 hr was obtained. In experiments with $HCN:H_2O:H_2$ or $HCN:O_2:H_2$ mixtures with ratios $1:1:100$, only approximately half the carbon and nitrogen have been accounted for by volatile products. This phenomenon is being investigated.

Experiments to determine the temperature dependence of reactions pertinent to this problem are now under way. In particular, the $HCN-H_2$ mixture is now being studied at 1200 K. The experimental procedure is being revised to eliminate problems that have arisen during the course of long-duration, high-temperature experiments.

Activation energies for molecular reactions of interest are generally high, 50 to 100 kcal/mole. Assuming 50 kcal/mole for the $HCN-H_2$ reaction, half-lives as a function of temperature are given below:

T (K)	τ_{e-1}
1093	18 hr
1000	6 days
900	100 days
800	9.2 yr
700	800 yr
600	3×10^5 yr

With a dynamical time scale measured in days, e.g., for mass ejection from stellar atmospheres, equilibrium would not be obtained for temperatures below 1000 K. For time scales less than 100 yr, as in Cameron's latest models of the primordial solar nebula, the equilibrium would begin to freeze at 700 K. An additional feature to be noted is the effect of a nonequilibrium excess of atoms or radicals. Because of the low activation energies of radical-molecule reactions, these processes may determine the composition in spite of their much lower concentration. An extensive laboratory investigation of chemical kinetics and molecular equilibrium is necessary in order to estimate reliably the composition of cosmic clouds.

AMINO ACIDS IN METEORITES

J. G. LAWLESS, E. PETERSON, and K. A. KVENVOLDEN

NASA/Ames Research Center, Moffett Field, Calif. 94035, U.S.A.

Since the initial finding of both protein and nonprotein amino acids with nearly equal abundances of D and L isomers of individual amino acids in the Murchison meteorite (a type II carbonaceous chondrite) [1], continued investigations of this and other meteorites (Murray, a type II carbonaceous chondrite, and Orgueil, a type I carbonaceous chondrite) have substantiated the early results [2], These findings have prompted a more detailed study of the compounds present in extracts of the Murchison meteorite. In addition to the 18 amino acids identified in earlier work [1, 2], there are at least 17 others. This population consists of a wide variety of polyfunctional, cyclic, and linear amino acids. Generally, the concentration of amino acids within this population decreases as the number of carbon atoms in the molecule increases. Studies show that the suite of amino acids found initially is very similar to that found in a completely different stone of this same fall, suggesting that the observed compounds are common throughout the meteorite. In an effort to understand the mode of occurrence of amino acids, a sample of this second fragment was treated with deuterated reagents. The sample was extracted with D_2O, and the N-trifluoroacetyl-D-2butyl esters were prepared to facilitate analysis by gas chromatography and gas chromatography combined with mass spectrometry. No deuterium incorporation was noted in any of the amino acids present. A portion of this same D_2O extract was hydrolyzed with 6N DCl in D_2O. Some incorporation of deuterium was noted in the amino acids recovered from this hydrolysate.

References

1. Kvenvolden, K. A., Lawless, J. G., Pering, K., Peterson, E., Flores, J., Ponnamperuma, C., Kaplan, I. R., and Moore, C.: *Nature* **228**, 923 (1970).
2. Kvenvolden, K. A., Lawless, J. G., and Ponnamperuma, C.: *Proc. Nat. Acad. Sci.* **68**, 486 (1971); Cronin, J. R. and, Moore, C. B.: *Science*, **172**, 1327, (1971); Oro, J., Gibert, J., Lichtenstein, H., Wikstrom, S., and Flory, D. A.: *Nature* **230**, 105 (1971); Lawless, J. G., Kvenvolden, K. A., Peterson, E., Ponnamperuma, C., and Moore, C.: *Science* **173**, 626 (1971); Lawless, J. G., Kvenvolden, K. A., Peterson, E., Ponnamperuma, C., and Jarosewich, E.: *Nature* **236**, 66 (1972).

URANIUM AND THORIUM MICRODISTRIBUTIONS IN METEORITES

G. CROZAZ*, D. BURNETT, and R. WALKER**

Division of Geological and Planetary Sciences, California Institute of Technology, Pasadena, Calif. 91109, U.S.A.

Galactic chronologies based on $^{244}Pu/^{238}U$ ratios in meteorites have assumed that there is no geochemical segregation between uranium and plutonium. In a check on the possibility of chemical fractionation, the microdistribution of uranium and thorium between various meteoritic phases has been studied in 15 meteorites (6 calcium-rich achondrites, 6 chondrites, 1 mesosiderite, and 2 iron meteorites). It is likely that if uranium and thorium are fractionated, plutonium and uranium will also be. The uranium-rich phases have been located by the fission-track method and identified with a microprobe. The uranium is generally concentrated in phosphates – usually whitlockite and/or chlorapatite – and in one instance (Stannern), in fluorapatite. The uranium concentration in a given phosphate phase is almost constant for a given meteorite and highly variable from meteorite to meteorite (0.2 to $>$ 10 ppm). Uranium contents of 200 to 300 ppb have been found in some achondritic pyroxenes.

A method to determine the thorium microdistribution on polished sections of the same meteorite has been developed. It is based on the observation of the fission-track distribution after irradiation with fast neutrons. For typical (Th/U $\sim$ 4) grains, essentially equal fission-track densities are observed from ^{232}Th and ^{238}U. Our techniques permit a quantitative measurement of thorium contents of $\sim$ 1 ppm in individual grains of 100 μ or larger. Preliminary results show that Th/U $\sim$ 4 in augite from Angra dos Reis and $\sim$ 10 in chlorapatite from Nakhla chlorapatite, and probably on total rock samples, would give anomalously high ratios.

SOME CONSTRAINTS ON CURRENT OUTGASSING OF WATER VAPOR ON MARS

S. J. PEALE

Dept. of Physics, University of California, Santa Barbara, Calif. 93106, U.S.A.

Leovy et al. [1] have offered a qualified suggestion that the diurnal brightening of the W cloud region on the surface of Mars may be due to daily outgassing of water vapor from a thermally active interior source. A lower bound on the areal density of atmospheric water vapor necessary for the observed brightening is determined to be comparable to the measured precipitable water over the planet for a reasonable range of hypothetical ice-cloud altitudes. A daily outgassing of this water vapor over the brightened region does not appear to be in conflict with observations, provided that the sources are

* On leave from Université Libre de Bruxelles, Belgium.
** On leave from Washington University, St. Louis, Missouri.

well distributed over the region and that the outgassing has persisted only over a restricted time period. On the other hand, daily uplift of atmospheric water from lower altitudes to the high plateau below the W cloud, where it saturates, is perhaps a more reasonable explanation that requires no local outgassing at all.

$^{40}Ar/^{39}Ar$ AGES OF LUNAR SAMPLES FROM HADLEY RILLE AND THE APENNINE FRONT

L. HUSAIN

Dept. of Earth and Space Sciences, State University of New York, Stony Brook, N.Y. 11790, U.S.A.

Gas-retention ages, cosmic-ray exposure ages, and the abundance of trapped argon have been determined for a suite of lunar samples from seven stations at the Apollo 15 landing site. The samples include fragments from 5 large crystalline rocks, 8 walnut-sized basalts weighing less than 40 g, 40 2- to 4-mm 'coarse-fines' soil fragments (including 13 basalts and 27 breccias), splash glass from 15286 and 15465, and 'green clod' 15426. The samples were dated by the $^{40}Ar/^{39}Ar$ method using stepwise heating. The results have been correlated with mineralogic-petrologic studies of these samples. Samples representative of all major petrographic types so far described have been examined. The gas-retention ages of the basalts range from 3.1 to 3.4 G.y., although most cluster around 3.3 G.y. The differences in basalt ages are outside statistical errors, although no correlation between the gas-retention ages and the basalt petrographic types has been observed. No basalts younger than 3.1 G.y. were found. Because of the large number of samples studies, it is inferred that no significant volcanic activity in this part of the Moon has occurred in the last 3.1. G.y.

Two recrystallized breccia fragments from the Apennine Front give minimum K/Ar ages of 3.96 ± 0.02 and 3.90 ± 0.02 G.y. The gas-release patterns of these breccias indicate a complex thermal history. The other breccias could not be dated, because they possess large excesses of extraneous trapped argon. Splash-gass fragment 15286, 10 and surface glass from breccia 15465 give complex release patterns and minimum ages of 0.99 ± 0.24 and 1.09 ± 0.14 G.y., respectively. This may be the age of the Aristillus and/or Autolycus cratering events. However, until substantiated by other similar glass fragments, this age should be taken with caution. The mafic green-glass spherules of unique chemical composition and the yellow-green, partially devitrified spherules from 15426 both give ages of 3.88 ± 0.07 and 3.80 ± 0.10 G.y. These samples contain inherent argon as well as solar-wind ^{36}Ar, and the measurements need further refinement. The cosmic-ray exposure ages of all samples range from 30 to 600 m.y.

LUNAR CHRONOLOGY BASED ON $^{40}Ar/^{39}Ar$ AGES

O. A. SCHAEFFER

Dept. of Earth and Space Sceinces, State Unveirsity of New York, Stony Brook, N.Y. 11790, U.S.A.

The $^{40}Ar/^{39}Ar$ method of age determination is particularly suited to lunar samples. The small sample size required (in the milligram range) allows the dating of numerous rock clasts in breccias as well as in the lunar soil. As the ages are relative to a standard of known $^{40}Ar/^{40}K$ ratio, it is of interest to compare them to a given standard to determine the relatively low age difference present on the moon. The $^{40}Ar/^{39}Ar$ gas-retention ages determined at Stony Brook are summarized. The ages have been determined for large crystalline rocks, small fragments from the soil, and breccias for Apollo missions 11, 12, 14, 15, and 16. The mare-type basalts all lie in the range 3.1 to 3.7×10^9 yr. The pre-mare material at the Apennine Front and in the Fra Mauro formation is as old as 4.1×10^9 yr. Of particular importance to lunar chronology is the age of the Imbrium event. By a combination of stratigraphy and absolute ages, it is possible to date this event as $3.75 \pm 0.05 \times 10^9$ yr. The ages of maria appear to be in accord with Baldwin's original suggestion that the Imbrium event released basalt from the lunar interior that flooded the other mare regions. The cosmic-ray exposure-age determinations of lunar rocks help elucidate the gardening of the lunar surface and the formation of the lunar regolith.

HIGH-SENSITIVITY AND HIGH-ACCURACY ANALYSIS OF RARE-EARTH ELEMENTS IN LUNAR ROCKS

L.-D. NGUYEN, G. PUIL, M. DE SAINT SIMON, and Y. YOKOYAMA

Centre des Faibles Radioactivités, Centre National de la Recherche Scientifique,
91 Gif-sur-Yvette, France

The aim of our work is to perform with high accuracy and sensitivity the analysis of rare-Earth elements contained in very small quantities of samples (1 mg or less) in order to investigate the variation of their concentration in different mineral grains of lunar soils. New methods are elaborated that improve greatly the isotopic dilution techniques with surface-ionization mass spectrometry.

Reducing pollution by chemical separation. In all our chemical processes, we used only 0.3 ml of solution to obtain all the rare-Earth elements in one fraction.

Increasing the accuracy of spike calibrations. Spikes are calibrated against standard solutions prepared from rare-Earth oxides of normal isotopic abundance. All our products are submitted before use to thermal-balance analysis to ensure anionic purity and stoichiometry. [1]

Increasing the accuracy of identifying rare-Earth elements. We use a method based on our systematic study of the decomposition of rare-Earth oxides on a heated rhenium filament to identify different rare-Earth elements. [2]

Increasing the sensitivity of the mass spectrometer. By means of ion counting, our detection system allows us to detect less than 1 ion/min (i.e., a current of 10^{-21} A). [3]

With these conditions, we performed analyses of rare-Earth elements contained in 1 mg of BCRP (USGS). The experimental results are in good agreement with those obtained by other investigators with greater quantities of samples. Lunar-sample analyses are being investigated.

References

1. Nguyen, L.-D., Puil, G., de Saint Simon, M., Gerdanian, P., and Yokoyama, Y.: to be published.
2. Nguyen, L.-D. and de Saint Simon, M.: *Int. J. Mass Spectr. Ion Phys.* **9** (1972).
3. Nguyen, L.-D., Goby, G., and Rosenbaum, B.: to be published.

COMPARATIVE LEAD-ISOTOPE DATING OF DIFFERENT CLASSES OF METEORITES

J. M. HUEY and T. P. KOHMAN

Dept. of Chemistry, Carnegie-Mellon University, Pittsburgh, Pa. 15213, U.S.A.

The isotopic composition of lead has been measured in 16 chondrites and 1 achondrite, and the lead contents have been determined by isotope dilution. The best least-squares isochron for the chondrites is $(^{207}Pb/^{204}Pb) = (4.718 \pm 0.001) + (0.5991 \pm 0.0032)(^{206}Pb/^{204}Pb)$, corresponding to a mean formation age of 4.505 ± 0.008 G.y. (using the most recent half-life values). Individual chondrites yield distinctly different ages, with a range of ~ 50 m.y. about the mean, but this may be due in part to variations in initial lead.

The Norton Country achondrite point does not lie on the isochron for chondrites and yields an age of 4.550 ± 0.003 G.y., confirming Rb/Sr indications that achondrites may have been formed ~ 50 m.y. earlier than the chondrites. The chondrite isochron does not pass through the currently accepted value for primordial lead, based on Canyon Diablo troilite, indicating the possible necessity for a more complicated evolutionary model, with the chondrites forming ~ 50 m.y. later than that iron meteorite.

The $^{207}Pb/^{206}Pb$ method is potentially capable of great absolute accuracy and fine time resolution. Additional meteorite measurements in progress should confirm or refute the indications of different formation times of different meteorite classes and of different meteorites of the same class.

THE PROBLEM OF COMPOSITION STUDIES OF
SOLID COMETARY MATERIAL

P. M. MILLMAN

National Research Council of Canada, Ottawa, Canada

Solid cometary material identified with certainty has not yet become available for laboratory analysis. Spectroscopic observations of radiating atoms and molecules originating in comets are available from two sources, both difficult to interpret in terms of quantitative elemental abundances. In the case of comet spectra, the material radiates in the environment of the solar wind and in a flux of intense solar electromagnetic radiation. In the case of meteor spectra, the material radiates in the environment of a strong effective wind in the terrestrial atmosphere.

Current studies in the laboratory of atomic- and molecular-collision cross sections, under conditions simulating the second environment, give promise of new information on the chemical composition of cometary fragments. The present state of progress in this field is summarized in relation to different groups of cometary meteoroids.

INTERPLANETARY DUST: A SOURCE OF PRIMITIVE MATTER

D. E. BROWNLEE and P. W. HODGE

Astronomy Dept., University of Washington, Seattle, Wash. 98195, U.S.A.

Trace elements from studies of lunar fines and analysis of meteor trajectories suggest that the bulk of submillimeter interplanetary particles are cometary with unfractionated solar abundances. Recently, samples of this material have become available for cosmochemical investigations. A very sensitive particle-collecting device flown in the stratosphere has collected particles with chondritic compositions that appear to be both true micrometeorites and debris of bodies that ablated or fragmented in the mesosphere. These particles are of considerable interest because they are a source of possibly primordial matter that has not previously been investigated. Because the particles are small, however, sensitive analysis is difficult. Analysis techniques and possible information on early processes and conditions in the solar nebula will be discussed.

ANOMALOUS GEOCHEMICAL COMPOSITION IN SPHERULES
AS A CRITERION OF COSMIC ORIGIN

T. GRJEBINE

*Centre des Faibles Radioactivités, Centre National de la Recherche Scientifique
91 Gif-sur-Yvette, France*

In the last 10 yr, a large number of microscopic spherules collected in the stratosphere or in polar ice have been analyzed. More than 500 analyses have been used for this review.

The average composition and percentage of different elements compared to silicon or iron do not show a terrestrial pattern, and the elements appear to be much less segregated than they are on the Earth, on the Moon, or in chondrites.

On the contrary, the composition, particle by particle, shows very diversified patterns, but these variations are quite different from the usual terrestrial mineralogical variations.

The main terrestrial, segregated groups are absent. Silicon in the form of quartz is also absent, and there are practically no spherules that show feldspathic composition. Whereas on the Earth most minerals have at least a small amount of aluminum, a large number of spherules show an association of silicon with metals without any aluminum. Another large group appears among spherules without any similar terrestrial associations; its composition shows a mixture of sulphur and chlorine with silico-aluminate.

The variety of composition is very large; only 3.5 spherules show a similar composition. The great terrestrial differences (lithophile, siderophile, or chalcophile segregation) are not respected.

Analyses of volcanic ash collected *during eruption and on the spot*, on the other hand, show a pattern of association of elements very similar to the rest of terrestrial mineralogy, and no confusion can be made with the two types of materials.

The anomalous geochemical composition of spherules must be related to a very special manner of formation.

MAGNETIC FIELD IN THE PRIMORDIAL SOLAR NEBULA

M. W. ROWE and J. M. HERNDON

Dept. of Chemistry, Texas A and M University, College Station, Tex. 77843, U.S.A.

D. E. WATSON

*National Oceanic and Atmospheric Administration, Earth Science Laboratory
Boulder, Colo. 80702, U.S.A.*

and

E. E. LARSON

Dept. of Geology, University of Colorado, Boulder, Colo. 80302, U.S.A.

A natural remnant magnetization of the Murray carbonaceous chondrite has been studied in detail. We have found evidence indicating the existence of a magnetic field in the very early history of the solar system – one that predates the accretion of the Murray material into a solid body. This field presumably existed during the condensation of the primordial solar nebula. A magnetic field was later operating on the accreted sample as the Murray parent body was heated to $\sim 450\,°C$.

THE EUROPIUM ANOMALY AND RARE-EARTH AND OTHER ABUNDANCES IN CALCIUM-POOR ACHONDRITES

W. V. BOYNTON and R. A. SCHMITT

*Dept. of Chemistry and the Radiation Center, Oregon State University,
Corvallis, Or. 97331, U.S.A.*

Twelve calcium-poor achondrites have been studied via instrumental neutron-activation analysis: five enstatite, four hypersthene, one olivine, and two olivine-pigeonite achondrites. Europium was found to be depleted in four enstatite and three hypersthene achondrites, which confirms the previous observations of Schmitt *et al* [1]. The Sm/Eu ratios in the samples depleted in europium range from 4.1 to 5.7 in the enstatite achondrites and from 6 to 13 in the hypersthene achondrites. This ratio compares with 2.76 for chondrites and a range of 3 to 12 for lunar soils and rocks. In every sample in which a europium depletion was observed, a depletion of the light REE relative to the heavy REE was also observed. This trend is consistent with REE distributions in pyroxenes. In one enstatite achondrite, Bishopville, a europium enrichment was observed (Sm/Eu=1.8), and there was no fractionation between light and heavy REE. The dark phase of Cumberland Falls yielded no europium anomaly, but the heavy REE were enriched by a factor of 2 relative to the light REE. Abundances of lanthanum in enstatite and hypersthene achondrites range from 0.1 to 2.4 times the average chondritic lanthanum abundance. The REE abundances in the olivine-pigeonite achondrites could not be determined except for lanthanum and europium, which are depleted by a factor of ≈ 25 relative to chondrites. The europium anomaly observed in most calcium-poor achondrites suggests that the reducing environment was similar for the formation of lunar basaltic rocks and calcium-poor achondrites.

Chromium and vanadium were found to be correlated in both the enstatite and hypersthene achondrites. The V/Cr ratio of 0.18 is the same as in chlorine chondrites and howardites but is ≈ 3 times less than the ratio in lunar material and ≈ 50 times less than the ratios found in oceanic basalts [2]. Scandium, iron, and manganese, which are strongly correlated in lunar material, are not correlated in the calcium-poor achondrites. The significant differences between the V/Cr ratios in oceanic and lunar basalts and calcium-poor achondrites and the lack of correlations between Fe-Sc and Fe-Mn suggest different parent matter for the Earth, Moon, and asteroid bodies.

References

1. Schmitt, R. A., Smith, R. H., and Olehy, D. A.: *Geochim. Cosmochim. Acta* **27**, 1077 (1963).
2. Laul, J. C., Wakita, H. W., Showalter, D. L., Boynton, W. V., and Schmitt, R. A.: *Proc. Third Lunar Sci. Conf.* (1972).

A SEARCH FOR INTERSTELLAR LITHIUM

N. CARLETON and W. TRAUB

Smithsonian Astrophysical Observatory and Harvard University, Cambridge, Mass. 02138, U.S.A.

We have recently made an intensive search for interstellar lithium at 6708 Å in the spectrum of ξ Oph. We used a high-resolution Pepsios interferometer with a full-width of about 100 mÅ and obtained about 2×10^6 counts per resolution element. Preliminary data analysis suggests an upper limit of 0.5 mÅ for the equivalent width of the lithium feature. It is anticipated that further analysis of these data could increase our sensitivity by about a factor of 3.

ASTROPHYSICS AND SPACE SCIENCE LIBRARY

Edited by

J. E. Blamont, R. L. F. Boyd, L. Goldberg, C. de Jager, Z. Kopal, G. H. Ludwig, R. Lüst,
B. M. McCormac, H. E. Newell, L. I. Sedov, Z. Švestka, and W. de Graaff

1. C. de Jager (ed.), *The Solar Spectrum. Proceedings of the Symposium held at the University of Utrecht, 26–31 August, 1963.* 1965, XIV + 417 pp.
2. J. Ortner and H. Maseland (eds.), *Introduction to Solar Terrestrial Relations. Proceedings of the Summer School in Space Physics held in Alpbach, Austria, July 15–August 10, 1963 and Organized by the European Preparatory Commission for Space Research.* 1965, IX + 506 pp.
3. C. C. Chang and S. S. Huang (eds.), *Proceedings of the Plasma Space Science Symposium, Held at the Catholic University of America, Washington, D.C., June 11–14, 1963.* 1965, IX + 377 pp.
4. Zdeněk Kopal, *An Introduction to the Study of the Moon.* 1966, XII + 464 pp.
5. Billy M. McCormac (ed.), *Radiation Trapped in the Earth's Magnetic Field. Proceedings of the Advanced Study Institute, Held at the Chr. Michelsen Institute, Bergen, Norway, August 16–September 3, 1965,* XII + 901 pp.
6. A. B. Underhill, *The Early Type Stars.* 1966. XIII + 282 pp.
7. Jean Kovalevsky, *Introduction to Celestial Mechanics,* 1967, VIII + 427 pp.
8. Zdeněk Kopal and Constantine L. Goudas (eds.), *Measure of the Moon. Proceedings of the Second International Conference on Selenodesy and Lunar Topography held in the University of Manchester, England, May 30–June 4, 1966.* 1967, XVIII + 479 pp.
9. J. G. Emming (ed.), *Electromagnetic Radiation in Space. Proceedings of the Third ESRO Summer School in Space Physics, held in Alpbach, Austria, from 19 July to 13 August, 1965.* 1968, VIII + 307 pp.
10. R. L. Carovillano, John F. McClay, and Henry R. Radoski (eds.), *Physics of the Magnetosphere. Based upon the Proceedings of the Conference held at Boston College, June 19–28, 1967.* 1968, X + 686 pp.
11. Syun-Ichi Akasofu, *Polar and Magnetospheric Substorms.* 1968, XVIII + 280 pp.
12. Peter M. Millman (ed.), *Meteorite Research. Proceedings of a Symposium on Meteorite Research held in Vienna, Austria, 7–13 August, 1968.* 1969, XV + 941 pp.
13. Margherita Hack (ed.), *Mass Loss from Stars. Proceedings of the Second Trieste Colloquium on Astrophysics, 12–17 September, 1968.* 1969, XII + 345 pp.
14. N. D'Angelo (ed.), *Low-Frequency Waves and Irregularities in the Ionosphere. Proceedings of the 2nd ESRIN-ESLAB Symposium, held in Frascati, Italy, 23–27 September, 1968.* 1969, VII + 218 pp.
15. G. A. Partel (ed.), *Space Engineering. Proceedings of the Second International Conference on Space Engineering, held at the Fondazione Giorgio Cini, Isola di San Giorgio, Venice, Italy, May 7–10, 1969.* 1970, XI + 728 pp.
16. S. Fred Singer (ed.), *Manned Laboratories in Space. Second International Orbital Laboratory Symposium.* 1969, XIII + 133 pp.
17. B. M. McCormac (ed.), *Particles and Fields in the Magnetosphere. Symposium Organized by the Summer Advanced Study Institute, held at the University of California, Santa Barbara, Calif. August 4–15, 1969.* 1970, XI + 450 pp.
18. Jean-Claude Pecker, *Experimental Astronomy.* 1970, X + 105 pp.
19. V. Manno and D. E. Page (eds.), *Intercorrelated Satellite Observations related to Solar Events. Proceedings of the Third ESLAB/ESRIN Symposium held in Noordwijk, The Netherlands, September 16–19, 1969.* 1970, XVI + 627 pp.
20. L. Mansinha, D. E. Smylie and A. E. Beck, *Earthquake Displacement Fields and the Rotation of the Earth. A NATO Advanced Study Institute Conference Organized by the Department of Geophysics, University of Western Ontario, London, Canada, June 22–28, 1969.* 1970, XI + 308 pp.
21. Jean-Claude Pecker, *Space Observatories.* 1970, XI + 120 pp.

22. L. N. Mavridis (ed.), *Structure and Evolution of the Galaxy, Proceedings of the Nato Advanced Study Institute, held in Athens, September 3–19, 1969.* 1971, VII + 312 pp.
23. A. Muller (ed.), *The Magellanic Clouds. A European Southern Observatory Presentation: Principal Prospects, Current Observational and Theoretical Approaches' and Prospects for Future Research. Based on the Symposium on the Magellanic Clouds, held in Santiago de Chile, March 1969, on the Occasion of the Dedication of the European Southern Observatory.* 1971, XII + 189 pp.
24. B. M. McCormac (ed.), *The Radiating Atmosphere. Proceedings of a Symposium Organized by the Summer Advanced Study Institute, held at Queen's University, Kingston, Ontario, August 3–14, 1970.* 1971, XI + 455 pp.
25. G. Fiocco (ed.), *Mesospheric Models and Related Experiments. Proceedings of the 4th ESRIN-ESLAB Symposium, held at Frascati, Italy, July 6–10, 1970.* 1971, VIII + 298 pp.
26. I. Atanasijević, *Selected Exercises in Galactic Astronomy.* 1971, XII + 144 pp.
27. C. J. Macris (ed.), *Physics of the Solar Corona. Proceedings of NATO Advanced Study Institute on Physics of the Solar Corona, held at Cavouri-Vouliagmeni, Athens, Greece, 6–17 September 1970.* 1971, XII + 345 pp.
28. F. Delobeau, *The Environment of the Earth.* 1971, IX + 113 pp.
29. E. R. Dyer (general ed.), *Solar-Terrestrial Physics 1970. Proceedings of the International Symposium on Solar-Terrestrial Physics, held in Leningrad, U.S.S.R., 12–19 May 1970.* 1972, VIII + 938 pp.
30. V. Manno and J. Ring (eds.), *Infrared Detection Techniques for Space Research, Proceedings of the Fifth ESLAB-ESRIN Symposium held in Noordwijk, The Netherlands, June 8–11, 1971.* 1972. XII + 344 pp.
31. M. Lecar (ed.), *Gravitational N-Body Problem, Proceedings of IAU Colloquium No. 10, held in Cambridge, England, August 12–15, 1970.* 1972, XI + 441 pp.
32. B. M. McCormac (ed.), *Earth's Magnetospheric Processes. Proceedings of a Symposium Organized by the Summer Advanced Study Institute and Ninth ESRO Summer School, held in Cortina, Italy, August 30–September 10, 1971.* 1972, VIII + 417 pp.
33. Antonin Rükl, *Maps of Lunar Hemispheres.* 1972, V + 24 pp.
34. V. Kourganoff, *Introduction to the Physics of Stellar Interiors.* 1973, XI + 115 pp.
35. B. M. McCormac (ed.), *Physics and Chemistry of Upper Atmospheres.* 1973, VIII + 389 pp.
36. J. D. Fernie (ed.), *Variable Stars in Globular Clusters and in Related Systems.* 1973, VIII + 234 pp.
37. R. J. L. Grard (ed.), *Photon and Particle Interactions with Surfaces in Space (Proceedings of the 6th ESLAB Symposium, Noordwijk, The Netherlands).* 1973, XV + 577 pp.
38. Werner Israel (ed.), *Relativity Astrophysics and Cosmology (Proceedings of the 1972 BANFF Summer School).* 1973, IV + 322 pp.
39. B. O. Tapley and V. Szebehely (eds.), *Recent Advances in Dynamical Astronomy.* 1973, XIII + 468 pp.